Biodiversity

A Beginner's Guide

Beginners
GUIDES

Biodiversity
A Beginner's Guide

John Spicer

A Oneworld Book

First published by Oneworld Publications in 2006

This revised and updated edition first published in 2021

A CIP record for this title is available from the British Library

ISBN 978-0-86154-017-4
eISBN 978-0-86154-018-1

Typeset by Geethik Technologies
Printed and bound in Great Britain by Clays Ltd, Elcograf S.p.A.

Oneworld Publications
10 Bloomsbury Street, London WC1B 3SR, England

For Fiona, my fair one

Through history, the most important changes in society have come from the bottom up, from grassroots.

Greta Thunberg

Contents

Preface to the second edition

It is now fifteen years since *Biodiversity: A Beginner's Guide* was published. In the intervening period, in one sense so much has changed and at the same time little has changed. We are still losing and damaging biodiversity, but now faster than ever before – the pace of change has accelerated. Previously, to the nearest approximation every species was an insect. Now it is becoming clear that to the nearest approximation every 'species' is a microbe. The importance of viruses to biodiversity is clearer than it ever was in the past – what they are, how they work, and how they originate and evolve – and not just because of SARS-CoV-2 (the name of the virus, *not* a bacterium, whereas COVID-19 is the disease), although admittedly that's been difficult to miss. And while the general format of the book is roughly the same, it has had to change to accommodate not just new and updated material but new ways of thinking about that material and what it means.

And while the purpose of the book is essentially the same, to act as a beginner's guide to biodiversity, the way that aim is now approached is slightly different – more than ever I see this as a book hopefully to astound, but also to shock the senses, challenge thinking, and to whet the appetite for greater and deeper knowledge and understanding. That is why the reading list has been greatly expanded, to act as a gateway to the academic or further study of biodiversity, at school, university or at home, and as an introduction to the wealth of books that are already out there and to bridge the academic–popular divide.

My thanks to Mark Lord-Lear for reading and commenting on early versions of Chapter 6, and to Jon Bentley-Smith, of Oneworld, for his excellent editorial assistance, and for that greatest of editor-virtues, patience tinged with empathy. Thanks also to all who wrote to me suggesting changes, additions and corrections to the first edition – I have tried to make sure I've incorporated or addressed them in this new edition. As with the first edition, this new text will not always satisfy everyone, particularly those whose research speciality is an area of biodiversity that I barely touch on, don't include or, even worse, that they don't feel I do justice to. I don't believe any one person could produce a comprehensive book on all of biodiversity, even if we could agree on what biodiversity is in the first place. Instead I have tried to tell not *the*, but *a* story of biodiversity – what it is, where it is, how it got here, how we value it, how it's threatened, how it can be maintained – to tell it as a story that interests me, in the hope that it may interest others too.

Preface to the first edition

It is said that books are best written in community. Over the past fifteen years I have been extraordinarily fortunate in the scientists I have worked with or for. They have made a lasting impression on what I know and believe about biodiversity, and this book would not have been written without their input in so many ways. Thanks to Lorraine Maltby, Phil Warren, Dave Morritt and Kevin Gaston for providing such a stimulating environment in which to work and think when I was at the University of Sheffield, and Kevin in particular as he opened my eyes to the notion that biodiversity is a serious science. I feel privileged to have spent so much of my time at Sheffield discussing, investigating and writing with Kevin, and I thank him for allowing me to use the same broad outline for introducing novices to biodiversity that we came up with in the *Rising Sun* so many years ago.

I also thank my present colleagues, the members of the Marine Biology and Ecology Research Centre here at Plymouth – Rikka, Simon, Dave, the 'Bish', Kath, Mark, Martin, Pete, Andy, Paul, Miguel, Mal, Kerry, Jason and Steve – for their friendship and for making going into work on a Monday morning something to look forward to; all of the postgraduate students, postdoctoral fellows and academic staff with whom I have had the honour of adding just a little to our knowledge of what biodiversity is and how it works – Sally Marsh, Kirsten Richardson, Jeanette Sanders, David Johns, Tony Hawkins, Steve Widdicombe, Nick Hardman-Mountford, Mike Kendall, Nikki Dawdry, Jenny

Smirthwaite, Kate Arnold, Lucy Dando, Emily Hodgson, Anne Masson, Sanna Eriksson, Susie Pihl Baden, Jalle Strömberg, Peter Tiselius, Jenny Cowling, Jason Weeks, Andy Rees, Mona Mabrouk El-Gamal, the inimitable Dave Morritt, Alan Taylor, Andy Hill, Stuart Anderson, Warren Burggren, Roy Weber, Brian McMahon, Peter Duncan, Katherine Turner, Alistair Edwards, Peter Spencer Davies, Maria Thomasson, Bengt Liljebladh, Paul Bradley, Angela Raffo, Hayley Miles, Ula Janas, Hugh Tabel, Tim Blackburn and the inspirational Geoff Moore who, as well as co-supervising my doctorate, first opened up to me the wonder and science which characterises the best of biodiversity as an academic subject. I am grateful to Roger Byrne and Mick Uttley, two of the sharpest minds I've ever encountered, for their detailed feedback on the manuscript and Marsha Filion of Oneworld who also made helpful comments on the manuscript. Suffice to say none of the above are responsible for any errors, omissions, transgression or biases that remain. Although too numerous to mention by name, I certainly owe a large debt to all my undergraduate students who have taught me so much as they caught on to how exciting and threatened our biodiversity is.

I seem to have gone through a fair number of editors at Oneworld but my thanks is no less heartfelt to Victoria Roddam from that initial meeting in a coffee shop in Bath (of all places) where the whole project kicked off, Mark Hopwood who had to badger me for so long that (unrelatedly) he gave up work and went back to becoming a student of philosophy, and finally an even more long-suffering Mike Harpley and Marsha Filion. All have been marvellous in their patience and help.

Finally, thanks to Fiona, my wife, and my children, Ellie, Ethan and Ben, for being so understanding and supportive. Ben provided the artwork for the book, more than making up for my lack of ability in that area, and for that I am grateful. And it is to Fiona, my fair one, that I dedicate this book. As she well knows, none of this would have happened without her.

1

The pandemic of wounded biodiversity

> Looking back, you can usually find the moment of the birth of a new era, whereas, when it happened, it was one day hooked on to the tail of another.
>
> John Steinbeck, *Sweet Thursday*

Biodiversity – what was that again?

Until recently global climate change was never far from the news. There was talk of setting and reaching emission targets, and global school strikes, and international meetings to agree ways forward, all of them receiving good air time. Biodiversity was there too, but more in the background, belying its importance. However, at the beginning of 2020 both climate change and biodiversity seemed to disappear from the scene in the wake of the COVID-19 pandemic as the world struggled to work out how to respond to this *new* and global threat.

Even in the early months of the crisis, it was mooted that the appearance of SARS-CoV-2 was linked with biodiversity and particularly with biodiversity loss. Of all emerging infectious diseases in humans, 75% are transferred from animals to people – what we refer to as zoonotic diseases. Pathogens (micro-organisms that cause disease) are more likely to make that jump where there are changes in the environment, like deforestation, and when natural systems are under stress from human activity and climate change. In the midst of a pandemic that many suspected was linked to biodiversity loss, we had to face our ignorance of what the fundamental relationship between biodiversity and human health actually looked like. We are only beginning to form a detailed understanding of many aspects of biodiversity, including how life and our planet work, and the effects we are having on that working.

At the same time, tales of ecological recovery filled our TV screens and populated the newspapers – '"Nature is taking back Venice": wildlife returns to tourist-free city'. Some good news in a very dark time. Biodiversity bouncing back quickly when we're forced to let up on the stress we put it under.

What is now clear is that issues around what biodiversity is, what we are doing to it, and how we best maintain it are not going to go away, pandemic or not. Knowing about and understanding biodiversity is interesting in its own right. But more and more it is becoming clear that, no matter who you are, biodiversity matters – and because it matters it is worth knowing as much about it as possible and understanding it better, if we are to make lives for ourselves in the coming decades.

That said, trying to pin down exactly what we mean when we talk of biodiversity is challenging. It seems to mean different things to different people. Here is a word that many would agree is important to get to grips with, but which few of us, when we reflect on it, have a good handle on.

There are a number of reasonable definitions of biodiversity – over 80 of them in fact! Many have merit or offer a slightly different take on the notion. Fortunately, there is one definition that has gained international currency, signed up to by the 150 nations that put together the Convention on Biological Diversity at Rio de Janeiro, Brazil, in 1992. Here biodiversity was defined as 'the variability among living organisms from all sources including [among other things] terrestrial, marine and other aquatic ecosystems and the ecological complexes of which they are a part...[including] diversity within species, between species and of ecosystems'. In short, biodiversity is the variety of life – in all its manifestations. This sounds quite satisfying until we ask the million-dollar question. The question that reveals the extent to which the study of biodiversity is a science: how do we measure it? This is not so straightforward. And yet it goes to the very heart of what we mean when we talk about biodiversity or when we refer to the biodiversity of a particular area, country or region. A simple illustration will help us see where the challenges lie.

A long, leisurely trip to La Jolla

For as long as I can remember I have been fascinated by the seashore and the living creatures found there. Without doubt one of my favourite locations in the whole world (rivalling even the perfect beaches of my childhood on the west coast of Scotland) is the rocky shore at La Jolla, just north of San Diego, Southern California (SoCal) (Fig. 1). Bird Rock is a beach which gives its name to a small community at the north end of Pacific Beach and at the south end of La Jolla (Fig. 2). The coast that skirts the land was recognised as an area of special biological significance back in the 1970s with the designation of the San Diego–La Jolla Ecological Reserve. This reserve is now part of the larger La Jolla Marine Conservation Area and Bird Rock itself is within the South La Jolla State Marine Reserve, established less than a decade ago.

I first discovered Bird Rock beach and its intertidal life in 1994. Carefully negotiating the uneven shore around Bird Rock, following the outgoing tide and discovering the huge variety of marine life present, particularly in deep crevices and glistening tide pools, is an unforgettable experience. The biodiversity of

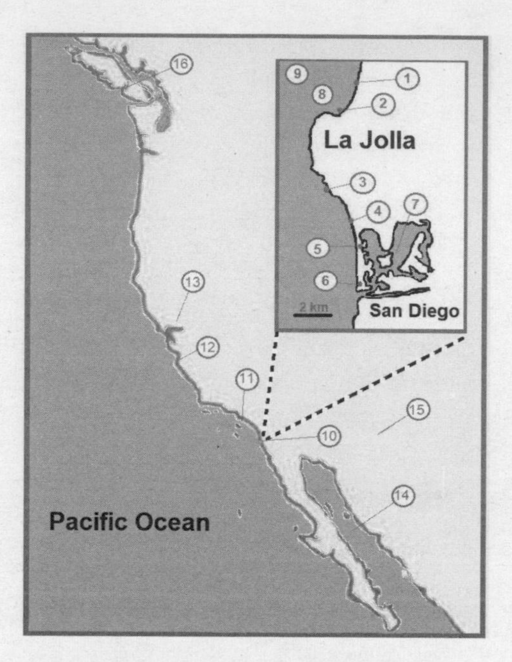

Figure 1 Location of Bird Rock in relation to other places in, or near, La Jolla (inset) and the north-east Pacific coast: 1) Scripps Institution of Oceanography, University of San Diego, 2) La Jolla Cove, 3) Bird Rock, 4) Pacific Beach, 5) Mission Beach north (the Mission restaurant), 6) Mission Beach south, 7) Mission Bay, 8) San Diego–La Jolla Ecological Reserve, 9) La Jolla deep-sea canyon, 10) San Diego, 11) Newport Beach, with the San Pedro Channel between there and Santa Catalina Island to the south, 12) Monterey, 13) Jepson Prairie, Solano County, 14) Gulf of California (Mexico), 15) Site of Biosphere 2, Oracle (Arizona), 16) Bamfield Marine Station, Vancouver Island (Canada).

Bird Rock is impressive, stunning even. As Ed Ricketts and Jack Calvin say in the preface to the first edition of their *Between Pacific Tides* (1939), 'our visitor to a rocky shore at low tide has entered possibly the most prolific zone in the world.' Between the tides (and maybe just above them), and in particular those that rise and fall on Bird Rock, nearly all the different major 'types' or 'designs'

Figure 2 Bird Rock with the tide receding (November 2018 © John Spicer).

of the things that characterise life on Earth can be found. Even nearby rocks contain spectacular fossils of relatives of the present-day squid, as well as clams, snails and lamp shells – the remains of ancient marine biodiversity that lived here about 80 million years (or Ma, an abbreviation we will use from now on) ago. Here we have 'life, and life in abundance'.

So let's return to where we started, 'How do we measure biodiversity?' and in this case let's say the biodiversity of Bird Rock specifically. How do we go about answering this question?

Could simply counting how many different types of living things there are do? This sounds like a possibility – but it's no mean task. It could potentially take not just weeks and months but years, maybe tens of years, even for such a small area – and that's leaving out all of the land animals, plants and microbes that form the larger landscape that is La Jolla. It is just about conceivable that for most of the largish marine animals, seaweeds and maritime plants we could put this list together. The California State Water Resources Control Board carried out a survey of life on the intertidal rocky shores within the San Diego–La Jolla Ecological Reserve in 1979. Close to Bird Rock they recorded fifty-three animal, nineteen seaweed and one flowering plant species, a total of seventy-three. This is a good start but still a huge underestimate of what is there, even based on my own visits. The marine life of these shores is, compared with similar beaches in other countries, relatively well known. This is largely because of the efforts of Ricketts and countless other marine biologists and naturalists, including those who have worked and studied (and still do) at the world-leading Scripps Institution of Oceanography, just north of Bird Rock, in La Jolla. However, for many microscopic animals, plants, fungi, bacteria and viruses our current information, while growing, is sketchy at best. It's not just the process of finding them that is problematic either. Many of these little specks of life are yet to be described, let alone the number of different types counted.

Even if it were possible to count the numbers of different types of living things, would such a list really be a measure of the biodiversity of Bird Rock? Well, perhaps. But it ignores the fact that there are rarely equal numbers of each of the living things present. Some organisms are extremely numerous and ubiquitous, while others are rare or only occur in particular, sometimes very localised, areas. In most cases, their numbers fluctuate throughout the year, or over even greater timescales. Periwinkles are absolutely everywhere, sometimes in large piles. Octopus can be found but there are not as many individuals as the periwinkles. Microbes may occur in tens of millions, outnumbering everything else many times over. Surely biodiversity must encompass not just differences but the actual numbers of different things present? So is it all about numbers or is it about difference – or is it both?

Up until now the differences we've considered are based on what an organism looks like and how that separates it from others. It doesn't take into account other differences that may be equally, maybe even more, important. Differences in genetics (the study of genes and their constituents, their heritability and variability) could be used to estimate how many things are there irrespective of whether they are big or small, identical in form, or difficult to tell apart. We could also use differences in the way individual types of organism 'work'. For example, how they acquire energy and what they do with that energy to maintain themselves. Also, what they actually contribute (if anything?) to the working of the living community to which they belong. We know that at Bird Rock limpets and sea urchins, two totally different types of animal, can do the same sort of thing – they do what cows do on land, they graze. Maybe taking into account what species do in an area is the key feature in any measure of biodiversity? And then there are the interrelationships between species. There are predators (the eaters) and there are prey (the eaten), for example. And what about the many thousands of parasites that live within the microbes, animals and seaweeds on the beach? In fact, there are a multitude of ways in which species and even groups of species influence other species or groups. And shouldn't the way living things work and interact, and the sum total of that – how an ecosystem functions – be more central to any measure of biodiversity we use? If we do take such a big-scale approach, what functions do we choose and why? And where do we factor in how visitors to the area interact with the living things on the beach, and how Bird Rock as a community of humans impacts on this small ecosystem, and how the beaches nearby, and the deep-water canyon nearby and all their inhabitants impact and influence the beach? And what about...and what about...and what about.

If you've been following all this, you may now find yourself at a mental crossroads. This is a well-trodden path. It is a place where many scientists, philosophers and theologians frequently find themselves. A place we will return to time and time again in the pages that follow. You can go down the 'oh, but the world's a complicated place' path: this grows into the 'we'll never get to grips with it' road, which leads to a comfy armchair, subdued lighting, a stiff drink, and an abandoning of intellectual pursuit and its travelling companion hope. Or you can opt for the 'okay, it's complicated, but...' path: a path where you may never find the *truth* (easily more difficult to define than biodiversity), but you are happy to settle for slightly less if it prevents you from stalling and keeps you walking, moving forward. The truth is there is no one way of measuring or quantifying biodiversity. We cannot measure the biodiversity of Bird Rock, or any other stretch of coastline, or of the ocean,* or of our planet for that matter. We can talk and think about the notion of biodiversity, but we cannot measure it – we can only measure *selected aspects* of it. Don't despair though. It may not be ideal – but even this is a good start.

Directions

To put together any beginner's guide to biodiversity, drawing on current scientific knowledge and understanding, much of our time will be spent looking at measures of biodiversity and how those measures change in time and space. Some measures will be better, or more appropriate, than others. In many cases we will find that the measure has been decided for us. Scientists often have to rely on the total number of species, the *species richness*, in a given geographical area just because that is the only information available, or likely to be available in the immediate future. Much work has gone into producing alternative measures. But given

---

* You'll notice that in this book we refer to the ocean, not the oceans plural. This is a recent change that is supposed to emphasise the unity of the water that covers our Earth.

the data we already have in scientific literature and museums, the relative ease of putting together inventories of different types of organism, particularly for very large areas, and the fact that it often reflects or incorporates numerous aspects of biodiversity that we've already discussed, species richness is not a bad measure. So much of what follows will use biodiversity and species richness almost as interchangeable terms – but not all the time.

There is no one way to write a beginner's guide to biodiversity. It could take the form of an all-singing, all-dancing panoply of the wonders and beauty of living things. It could be an encyclopaedic catalogue of the variety of living creatures and the places they live. It could focus on how to preserve biodiversity. Or it could combine aspects of all three with different emphases. So, what will be the approach here? It is an old, but I think insightful, idea that the only way you can ever say anything general and all-embracing is by starting with something tangible, specific, familiar, local. For example, over a hundred years ago 'Darwin's bulldog', Thomas Huxley, used the humble crayfish for the title and subject of a book he wrote to introduce interested readers to the general study of zoology. He used aspects of crayfish biology as illustrations, as jumping-off points to explore broader aspects of all animal life. In that same vein, throughout this book I will use the beautiful shores at La Jolla and aspects of my own experience, as a way into some of the big biodiversity issues and patterns. In that respect this is a very personal book.

What will be the key features of this beginner's guide? In the next chapter we ask: How many species are there currently on Earth and how are they distributed between the different large groupings of organisms we presently recognise? What are these large groupings and how have we ended up with them? This will involve asking another thorny question – what is a species anyway? In Chapter 3 we will see that biodiversity is not distributed evenly across Earth's surface. There are hotspots and there are coldspots. We will try to visualise the current patterns of biodiversity (or at least measures of that biodiversity), in particular how the number of organisms varies with latitude, altitude and depth. That should take us neatly on to the fourth chapter, where we delve briefly into the origins and development of biodiversity, concentrating particularly on the ups and downs of the past 600 million years. We'll enter into a debate on the origins of biodiversity that goes to the very centre of what we think about ourselves and the organisms with which we share this planet. Much of our attention will be on extinction, both past and present. We will briefly introduce what looks set to be a new geological era, the Anthropocene.

Up to this point in the book, biodiversity is discussed, at least as much as possible, as an objective body of scientific knowledge. But part of the reason we find it difficult to get a handle on the term is because, in the minds of many, it is a value-laden concept. Think about it. Because we have to rely on measures of biodiversity, and the measures we pick often reflect what it is we value about biodiversity, how could the whole subject not be value laden? So the remainder of the book is devoted to the threats to and value(s) of biodiversity, and in the case of the latter, including direct and indirect monetary value. The main direct (proximate) threats are introduced and illustrated before particular attention is paid to one of the main indirect (ultimate) drivers – us. We will consider attempts by economists to cost the Earth and its services (i.e. what it does for us); a project by scientists to create a living life-support system for eight people; a current scientific controversy on how many species we actually need; and a survey of philosophical and religious thought on the place of biodiversity, and nature in general, in our thoughts and beliefs. The penultimate chapter leads on from talk of value to what have we done, and what are we doing, to maintain biodiversity globally. We'll also ask the question, 'When it comes to maintaining biodiversity, what have we achieved?' The final chapter, for me, draws the whole book together, but in other ways it's optional. It is a personal view of what all of this biodiversity stuff means, and what we can do about it.

I've tried to make what follows quantitative rather than go for the *ooh-ah* factor. After all, the bread and butter of science is what you can measure or quantify in some way. That is not to say this is all that I personally value about biodiversity – facts and figures, calculations and guestimates. I have felt the wonder of peering down a microscope for hours on end watching an embryonic shrimp develop, the separation of its cells, witnessing its first heartbeat; experienced the numinous presence, the awe-filled delight, of trekking through the holy ground that is an ancient redwood forest. I am that calm, collected university professor who, with their students on a field trip, ends up being the object of bemused attention as he bellows, 'Look, there's a fin whale – it's a fin whale – oh!', just like a giddy school kid. But this is not the time, or the place. The majesty and wonder of biodiversity is always better 'felt than tell't'. But just now we live at a time in history when, I suggest, our living world will at best diminish and at worst change out of all recognition. It is good to try to get a handle on the facts of the matter, to help inform our thinking about and feel for biodiversity, and what we should do about it.

This beginner's guide to biodiversity is aimed mainly at those with little formal training in biology. It is for those who want to find a way into some of the most interesting biology questions, and some of the most pressing biodiversity issues, of our time. And be sure of one thing: If I am successful, this book is just a way in – I don't think it's any longer possible to produce a short, comprehensive introduction to biodiversity, a subject which, when you think about it, encompasses just about everything you can think of, and probably a lot more besides. For some readers such an introduction will be good enough. For others they will want to move on from this 'way in' to engage with more detailed or wider accounts and investigations of aspects of biodiversity. For yet others there will not be enough rigour here. They will want a more balanced, more detailed, more comprehensive (as I said, there's lots of stuff that many would consider essential that I don't even mention, let alone discuss), less personal account and discussion. Some may prefer a purely academic approach. For all those people I have provided a starter pack of key popular and academic books in the final 'Going further' section. The books are presented according to the chapters they are most relevant to. And to those who have already decided that the approach I'm proffering is not for them, I suggest putting the book down (having paid for it preferably), taking a deep breath to regain your composure, and having a look down the book list to find something that does light your candle. For those of you still with me, we'll start by asking, 'How many different living things are there on Earth and how are they related?'

2

Teeming boisterous life

The true biologist deals with life, with teeming boisterous life, and learns something from it.

John Steinbeck, *The Log from the Sea of Cortez*

The big picture

To think of life on our planet only in terms of facts and figures, percentages and ratios is, to many of us, simply *not good enough*. It's too narrow a view of *how things are*. And yet, if we are to come to some general consensus about the 'big picture' with regard to the living beings with which we share this planet, and work out what we are to make of them, we must attempt to stand back and consider the best facts available to us. We must try not to allow the many things that have been, and are, *believed* about life, about biodiversity, to cloud our judgement – not yet anyway.

In this chapter, and in the two that follow, we set about a short scientific exploration of life on Earth. What we want is to see clearly the big picture. Ironically, it's quite difficult to get this big picture – even from good biology textbooks and monographs. Biology courses at high school, and university too, often focus in on detail. And to such an extent that you need a thorough scientific training to follow, let alone attempt to understand, what is going on. So, is it possible to produce a big picture that can be read and understood by just anyone that is interested? I do hope so.

We'll start with a brief look at our planet itself and then consider in a bit more detail the living beings that inhabit it; how many different types there are, or have been, where they live and what they do.

The volleyball on Mission Beach

For most of us, getting to grips with things that are too big or too small to visualise is a challenge. How do we get some perspective on something as mega as *life on Earth*? Let's start where one of my sons lives and surfs, on Mission Beach, San Diego, SoCal, just a couple of kilometres south of Bird Rock (Fig. 1). Playing volleyball is big on this beach. Each properly inflated volleyball is about 22 cm in diameter. Imagine if we could shrink Earth to the size of this ball. Our planet is 12,756 km in diameter. The ball is a 1:58 million scale model. Just under a metre away imagine another ball about a third as big as the volleyball, also lying out on the sand. This represents the moon, 384,400 km from Earth. On this scale, the sun is a giant ball, over 24 m in diameter, sitting in spectacular fashion 36 minutes' walk, or 2.5 km, north along the boardwalk towards Pacific Beach (but stopping level with the amazing Mission Restaurant – a good idea whether the sun is shining or not). Managing to extricate yourself from the Mission, head north again. Next stop is the used-to-be-a-planet Pluto at the furthest reaches of our solar system. It would be an object, less than a quarter the diameter of the volleyball, sitting over 100 km away, or 28 hours' hike, just south of Newport Beach. The next nearest star, *Proxima centauri*, at 4.22 light years away (or 9 trillion km) would defeat the most ardent hiker as it would be one fifth of the way to the actual moon. For the next planet orbiting a sun like our own (exoplanet HD217014, orbiting 51 Pegasi) we're talking a fairly big ball twice as far as the actual moon from our volleyball on Mission Beach.

Returning to South Mission Beach volleyball courts, we pick up one of the volleyballs lying in the sand and look at it. Someone has written their name using a permanent marker. The ink mark is marginally thinner than the page you're reading. But it's still proportionally thicker than the space within which every living being can and does occur on Earth. All of life on Earth, and perhaps all life period, can be found within a band about 25 km thick across the Earth's surface (550 gigatons, or 550 billion metric tonnes, of carbon worth of life), half of one millimetre on the volleyball scale of things. The region where weather happens (the troposphere) is on average 12 km high, 10 km at the poles and 16 km at the equator. So between zero and ten kilometres high is where we find all flying and nearly all land-living life, and even then it is mostly concentrated at the bottom. I say nearly all land-living life because there is life found beneath the surface of the Earth, in caves, in soils or in rocks – in fact some very simple forms of microbial *deep life* have been recorded, and even thrive, as much as 6 km deep underground. Water covers just less than three quarters of Earth's surface. The average depth is 3.9 km with the deepest part being 11 km – and here too there is life. This thin band in which we find all the living beings on Earth (referred to as the biosphere) really is small and fragile in the great scheme of things. (But it does depend on what you mean by the 'scale of things': the mass of carbon in the biosphere is equivalent to the mass of 33,682,405 Greyhound buses, which if laid end-to-end would reach from here to the sun and back, and back to the sun again.)

Before we go on and look at how many types of living things there are, and what they are, we first have to ask the questions, 'Can all these types be referred to as species?' and, even more fundamentally, 'What do we mean when we say "this is a species"?'

'A rose by any other name'...what's a species?

For hundreds of years the species has been regarded as the common currency of life on Earth. The way we use the term is fundamental to how we've understood biodiversity, at least until fairly recently. It is difficult to talk about most areas within biology without having to refer to species in one way or another. And yet, as observed by an early evolutionist, Jean-Baptiste de Lamarck (1744–1829) in his *Recherches sur l'organisation des corps vivants*, 'The more our knowledge has advanced, the more our embarrassment increases when we attempt to define a species. The more natural specimens are collected, the more obvious it becomes that almost all gaps between species are filled and our dividing lines fade away.' No surprise then that there is still much debate over this seemingly most elementary of all labels. Currently there are at least seven different definitions of 'species', although here we'll only explore a few of those.

Morphological species

When most people use the word 'species', more often than not they are talking about the *morphological species*. Telling species apart has traditionally depended on noting differences in what organisms look like. For example, in the north-east Atlantic there are a number of fish that look like herring: the sardine, the sprat and the herring proper. They share a herring 'shape' but there are visible differences. And these differences, as long as they are consistent and if they breed true, can be used to separate the different species. For example, the lower jaw of both the herring and the sprat is consistently longer than the upper jaw, whereas the jaws of the sardine are of equal length. And the belly of the sprat has a distinct serrated edge and is strongly keeled compared with the herring, which is weakly keeled and not strongly serrated.

So biologists use a biological key to identify species. A key consists of morphological differences often organised into the type of question/answer approach that you, for instance, may have used to diagnose problems with, say, your washing machine. Start at first question – Is the 'On' light on? Yes – go to the next question. No – try plugging it in, stupid. Question two – Is the drum spinning? And so on. A simple key for discriminating between our three members of the herring lookalikes could be: Question 1 – Is the lower jaw the same length as the upper jaw? Yes – then it's a sardine. No – go to Question 2. So, Question 2 – Belly serrated? Yes – it's a sprat. No – it's a herring. Those of you who have seen a herring and a sardine may say, 'The herring's dark blue on top and the sardine's a sort of greenish-olive colour, is that not more obvious?' Well, yes, it is – as long as the creatures are alive or newly caught. But, as is the case with many biological collections, identification is often carried out on pickled specimens. This is often the case if there is a lot of stuff to identify. And the stuff used to pickle animals, formalin or alcohol, quickly removes colour and lifelike appearance. Choosing features that don't change after preservation such as relative lengths or numbers of structures is usually seen as the best option.

Identifying organisms in this way is both time-consuming and often requires specialist knowledge. The advent of iNaturalist, where you can take pictures on your smartphone and the image is identified for you, will potentially increase access to morphological species identification to anyone with a smartphone. However, often you still have to use a specialist key to get a definitive identification. There have been recent attempts to barcode living things to make the process of identification easier, quicker and more reliable. This approach still sort of relies on morphology. It uses differences in the microstructure of some of the common information-containing molecules of species – the genetic substances, DNA and RNA – to classify and identify the species they are extracted from. Such differences accumulate in each of two species (derived from the same ancestor) as the amount of time they are separated for increases. Use of genetic molecules in this way is seen by some as one of the best ways of identifying and cataloguing biodiversity, although it is still considered controversial, or at least premature, by others.

Identifying species without ever seeing them

An alternative approach to estimating the numbers and identities of different living things is by analysing environmental DNA (eDNA) using high-throughput DNA-sequencing technology. eDNA can be thought of as genetic material taken directly from the environment (water or soil, for example) but without knowing what the source material is. It is an approach that could be easily standardised, shouldn't have to rely on continuous input from experts in identification, should deal with the challenges of identifying 'hidden' (i.e. cryptic) species and juvenile stages, and it is non-invasive. Though still in its infancy, the technique shows promise for biodiversity work. It was first applied in analysis of sediments where DNA from animals and plants – some extinct, some still around today – was detected. It is the staple tool of microbiologists who regularly will replace 'species' with the less loaded term 'operational taxonomic unit' (or OTU). There are some major challenges ahead before the full potential of eDNA can be realised, centring on issues of contamination (and so generating false results) and the need to develop reference databases with which to compare your samples. Still – watch this space.

Biological species

Now, there are other biologists who think that even if two individuals are near identical in appearance, but live so far away from each other that there is no chance of them getting together (in a biblical way), then they should be considered separate species. So only an interbreeding bunch of neighbours would be referred to as a species, in this case a *biological species*. For example, Audubon's warbler, with its distinctive yellow throat, and the myrtle warbler, with its distinctive white throat, were originally described as two separate morphological species. However, it was later discovered that their breeding ranges overlapped in Alaska and north-west Canada, and here they successfully interbreed. So what were considered two species is now one, the yellow-rumped warbler. The biological species concept feels a step up from the morphological one, but its strength is also its weakness. It is reliant on sex. So what do we do with organisms that do not have sex, like little water fleas and many microbes?

Evolutionary species

A third definition of a species is the *evolutionary species*: a single bloodline of ancestor-descendants, with its own evolutionary path and history. To take another bird example, there are a number of different kinds of red crossbill that may occur together, but quite literally only mate with their own kind. Each of these different kinds would be an evolutionary species. The biological and evolutionary species are, together with the cohesion, ecological, phylogenetic and recognition species, all interesting takes on how to define a species. But the bottom line is that, at this moment in time, the morphological species is the easiest and cheapest one to use, and so will probably continue to be the most commonly used, at least for large-scale biodiversity studies.

But times they are a-changing. None of the species concepts covered above are sufficient to define a species of bacteria. Instead bacteria are recognised and named by a combination of how they look, how similar their genetics are, and how closely related they are. There is still a lot of uncertainty, and a lot of unknowns about bacterial 'species', or as some would prefer OTUs, but given their dominance (as we'll soon see), this is changing rapidly. So, again, watch this space.

Naming species

For the last couple of hundred years biologists have agreed amongst themselves to give species two-part scientific names (a binomen). These names are always latinised following the custom prior to the eighteenth century of publishing scientific papers in Latin. Latinised species names now provide a universal language for the naming of living creatures. So the blue whale has the scientific name *Balaenoptera musculus*.

It may come as a surprise to the non-specialist but no one actually regulates the names given to species, although some organisations and individuals do try. Often the name chosen reflects the imagination, understanding and/or perversity of the person(s) who decide upon it. For example, the narwhal is a whale with a huge horn protruding from its head and so is often referred to as the unicorn of the sea. Its scientific name is *Monodon monoceros*. Carolus Linnaeus (1707–78), the Swedish biologist who was first to suggest and use the binomen, originally described the narwhal as having 'one tooth, one horn'. The Latin name is constructed from three Greek words: *monos* = 'single', *odontos* = 'a tooth', and *keras* = 'a horn'. That works. However, not all species have such 'serious' names: *Agra vation* is the name of a tropical beetle that was very difficult to collect; *Crepidula fornicata* is a type of limpet with a very distinctive reproductive behaviour;

and *Massisteria marina* is a one-celled seawater-dweller that caused some commotion when it was first discovered. Finally, and much more personally, *Traskorchestianoetus spiceri* (Fig. 5.1f) is a tiny mite that inhabits the underbelly of beachfleas (Fig. 5.1b) living under seaweed, and amongst flotsam and jetsam, cast up at the top of the shore on North American beaches. A copy of the original species drawing was sent to me, as its discoverer, by Dr Matt Colloff, who described it in collaboration with a Belgian colleague. The framed drawing takes pride of place at the back of a small dark cupboard in my home.

How many living species...and what are they?

Exactly how many different types of species are alive on Earth today? To be truthful no one really knows because no one has gone out and counted them all. And at the moment there is still uncertainty on what it is we are actually counting. About ten years ago serious estimates varied between 3,635,000 and what was considered at the time a staggering 111,655,000, with the best estimate coming in at around 13.62 million species – still only a tenth of all species formally described. And this estimate was dominated by animal species, with the majority named being arthropods. Often when people talk or think of living things, they have in their minds people, dogs, cats, tigers, pandas, gerbils, maybe even lizards, frogs and toads and fish. These are all animals, and relatively large animals at that. But biodiversity is so much more than just large animals which, as we can see, could comprise less than 1% of the total number of species. Ironically, it has now become even more difficult to give an estimate of the number of species on Earth because of a small problem that isn't so small.

For many years, microbiologists have been trying to bring to popular attention the richness of species you cannot see with the naked eye. In 2017 Brendan Larsen and colleagues, incorporating new knowledge, particularly of microscopic life, and attempting to take a more rigorous approach to what we already knew, came to the conclusion that there are likely to be 1–6 billion living species on Earth, with 70–90% of them bacteria. If they are even slightly near the mark (see some cautionary comments a little further on), this fundamentally alters how we think of both the numbers and distribution of biodiversity – and, to 2017's nearest approximation, every species is a microbe (Fig. 3).

Figure 4 is a graphical representation of numbers of species, described and estimated for all but two of the major groups of life (the Bacteria and Archaea – the reasons for not including them here will become clear as we go on). So, let's go through each of the groupings presented on the graph, moving from most to least species rich. The groups are ranked according to numbers of described species, but as we shall see there are a couple of

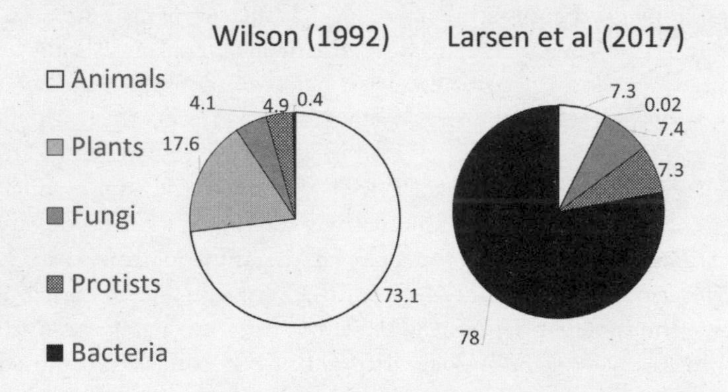

Figure 3 Traditional estimate of species richness based on Wilson's 1992 *The Diversity of Life* (left) compared with the latest estimate by Larsen and colleagues (2017) which takes recent assessments of microbial diversity into account (right). (Adapted from Larsen *et al.* (2017), *Quarterly Review of Biology* 92, 229–65.) The numbers given are percentages of the total number of species. Note that the plants don't make it onto the Larson pie chart as they only make up 0.02%.

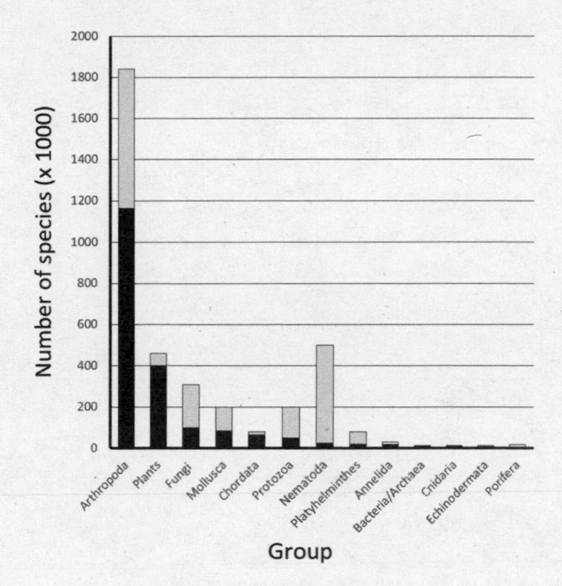

Figure 4 Graphical representation of numbers of species, described (black bars) and estimated total (black + open bars) for most of the major groups of life. Only described species are given for the Bacteria and Archaea, for reasons that will become clear. The groups are ranked according to numbers of described species, moving from most species rich (left) to least (right).

groupings which might upset this ranking if we substitute *predicted* for *described* numbers of species. To help us visualise, get a handle on, the different kinds or designs that make up nearly all of life on Earth, one could do worse than spend time exploring the Californian intertidal zone. So we return to Bird Rock in SoCal. The biodiversity we observe in this beautiful, though small, area of beach is, to some extent, a picture of what we find in other seas and even globally. Introducing the different groups of living organisms on, or around, this one beach gives us a way into introducing, illustrating and discussing life on a global scale (Fig. 5).

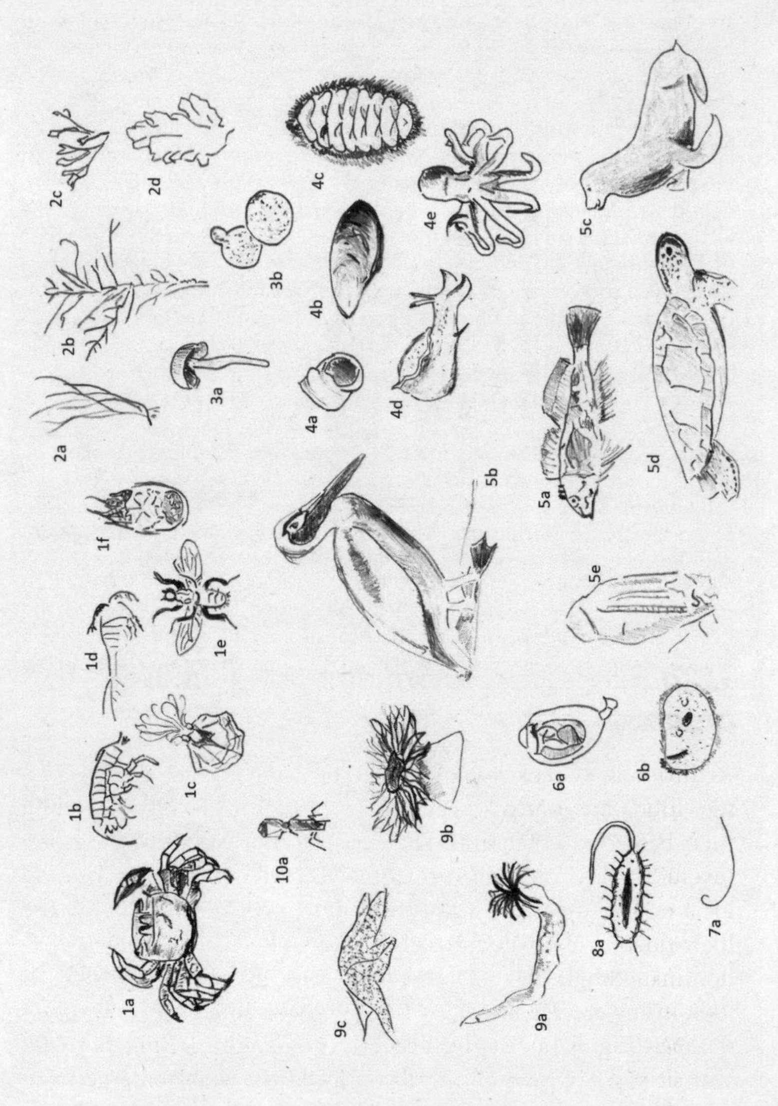
1a
1b
1c
1d
1e
1f
2a
2b
2c
2d
3a
3b
4a
4b
4c
4d
4e
5a
5b
5c
5d
5e
6a
6b
7a
8a
9a
9b
9c
10a

Figure 5 Representatives of the different groups that can be found on the beach at Bird Rock. Drawings are made from life, supplemented by various sources, courtesy of my daughter Ellie. 1a) striped shore crab *Pachygrapsus crassipes* (4 cm wide), 1b) beachflea *Traskorchestia traskiana* (1 cm long), 1c) barnacle *Balanus glandula* (2 cm wide), 1d) copepod *Tigriopus californicus* (2 mm long), 1e) kelp fly *Coelopa* sp., 1f) Histiostomatid mite (length 0.3 mm), 2a) surfgrass *Phyllospadix* (60 cm long), 2b) red seaweed *Corallina* (15 cm long), 2c) brown seaweed, channelled wrack *Silvetia*, formerly *Pelvetia* (90 cm long), 2d) common sea lettuce *Ulva* (30 cm long), 3a) mottlegill mushroom (5 cm tall), 3b) yeast (5 μm diameter), 4a) periwinkle *Littorina* (2 cm long), 4b) California mussel *Mytilus californianus* (13 cm long), 4c) woody chiton *Mopalia* (7 cm long), 4d) California sea hare *Aplysia californica* (25 cm long), 4e) two-spotted octopus *Octopus bimaculoides* (body = 20 cm long), 5a) woolly sculpin *Clinocottus analis* (18 cm long), 5b) brown pelican *Pelecanus occidentalis* (1.5 m long), 5c) Californian sea lion *Zalophus californianus* (2.5 m long), 5d) Pacific green turtle *Chelonia mydas* (1.5 m long), 5e) light bulb sea squirt *Clavelina* (1 cm long), 6a) sessile protozoan ciliate (30 μm long), 6b) mobile protozoan ciliate (80 μm long), 7a) free-living nematode worm (3 mm long), 8a) bacterium, 9a) polychaete worm *Hydroides* (2 cm long), 9b) sandy sea anemone *Anthopleura elegantissima* (8 cm wide), 9c) common seastar *Pisaster* (arm radius 15 cm), 10a) virus.

1) To the nearest approximation (almost) every organism is an arthropod...?

Arthropods (Greek *arthron* meaning 'joint', and *pous* meaning 'foot') are jointed-legged animals which grow by shedding their hard outer skeletons. They are the most species-rich group (excluding microbes) in and around Bird Rock, the beaches and local environment of La Jolla, and indeed worldwide. Below the high tidemark, the crustaceans, the so-called 'insects of the sea', dominate. High on the shore, turn over just about any rock, or look in any crevice, and there is the striped shore crab *Pachygrapsus crassipes* (Fig. 5.1a), or the beachflea *Traskorchestia* (Fig. 5.1b). If you sit still for long enough, you'll see the crabs emerge from their hiding places. Slightly further down the shore, there is the less pugnacious kelp crab *Pugettia*, one of a number of different types of crab on the shore. Search more carefully in the crevices, beneath rocks and in tide pools lower down, and you can find hermit crabs; stalked barnacles; acorn barnacles (Fig. 5.1c); porcelain crabs; pistol, red-rock, red-banded and slender green shrimps – an incredible array of relatives of the crabs that command our attention so easily. In the pools, beneath the rocks, in gravel and sand, and on and amongst the seaweeds, are many different types of microcrustaceans, mainly copepods (Fig. 5.1d) and seed-shrimps (ostracods) – but you would need a good hand lens, or even better a microscope, to see these.

There are an estimated 150,000 crustacean species worldwide, with only 47,000 currently described. However, this figure, though impressive, pales in comparison with their more land-based relatives. At Bird Rock the crustaceans may dominate below the high-tide level, but above and on land proper the insects and their relatives come into their own. Wrack flies (Fig. 5.1e) are found in a thick haze above the cast-up flotsam and jetsam, and many hundreds of species, beetles mainly, visit the wrack from nearby locations. As well as insects, it is possible to find representatives of the second most dominant terrestrial arthropod group, the arachnids. Spiders inhabit the strandline and nearby vegetation, but it is the mites (Fig. 5.1f) that are superabundant and can be found even on parts of the shore, or on animals and algae that are submerged during the high-tide period.

Until recently the land-living arthropods, the insects, arachnids, centipedes and millipedes, particularly when combined with the crustaceans, were considered to be the most species-rich group on Earth. With 1,165,320 species described, the total is estimated at 1,840 million, although some would place the figure as high as 8 million. Certainly, the accolade of the greatest number of described species goes to the insects – 1.02 million, and growing daily. They are properly waterproofed crustaceans if you like. Within the insects, the beetles are cited in many biology books as the most species-rich group on Earth. This is probably true but is still debated. Worldwide there are 102,248 described species, although it has been suggested that, to date, only one in ten arachnids are described.

The arthropods are of huge economic importance. The great majority of insects are herbivorous (plant-eating). They eat more plants than any other creature on the planet. So they include major pests of nearly every one of our major food crops (e.g. the Colorado potato beetle, which attacks potatoes, tomatoes and aubergines (eggplant if you're in North America), and the Yellow stem borer, which is the most pervasive insect pest of rice), as well as species that attack stored foods and the wood we use to build furniture, houses and the like. The costs of such pests can be astronomical. Take the current problem with termites which damage approximately 600,000 homes in the US each year. The cost of controlling the activities of the termites, and repairing the damage they do, is estimated at US$5 billion annually (2015 figures). This said, less than 1% of insects are pests, with only about 100 species being persistent nuisances. Other species feed on pests. Ladybirds feed on aphids, for example. And many crop species would not be able to reproduce without insects to help them by carrying pollen from one plant to another. Those species that live in the soil, together with those that scavenge, are critical in breaking down and recycling organic waste. There are some non-herbivorous insects, some of which (at best) are irritants to us and our domesticated animals thanks to their biting and sucking – those readers who initially thought themselves blessed to spend a calm summer evening beside a Scottish loch will be well acquainted with the *midge*, an extremely tiny, irritating, biting fly which can ruin a holiday in a matter of minutes.

At worst many of these biters and suckers spread disease: bubonic plague by fleas, and malaria and yellow fever by mosquitos are just a couple convincing examples. Some species of ant, beetle, caterpillar and locust are actually eaten by people, as are some insect products, such as honey from bees.

The arachnids contain many poisonous species of spiders and scorpions (some can be fatal), but perhaps the most important economically are the ticks and mites. Like some insects, they can parasitise, infect or just irritate people and their domestic animals. Many crustaceans such as crabs, lobsters and shrimps are an important food source, but we should not forget (if it was ever known in the first place) that it is many of the tiny microcrustaceans that form a vital component of freshwater and marine food webs and ecosystems.

2) Greenery: The Plantae

Now we have organisms that, unlike much of what has gone before, should be instantly recognisable, the plants, or more formally the Plantae. These are the mosses, ferns and other spore- and seed-bearing plants. In the inventory of organisms on the intertidal rocky shores of the San Diego–La Jolla Ecological Reserve only one flowering plant species is recorded in the survey. This is the ecologically important surfgrass, *Phyllospadix* (Fig. 5.2a). However, the report highlights that there is a small patch of true eel grass, *Zostera*, at a depth of 1 m. Also, around the high-water mark and on the cliffs and land adjacent to the beach, without really searching for them, I saw a wide variety of salt-tolerant flowering plants (succulents). So there are plants in and around the beach, but the beach is arguably not an ideal location if we really want to study plant diversity. (That said, it does depend what you mean by biodiversity because the differences between flowering plants that live in seawater and those that are 'merely' maritime is a fascinating study in its own: see, for example, Ian Hepburn's *Flowers of the Coast*.) Most plants are land-living and are found in forests, grasslands and even deserts, with very few returning to the water. In an area of 5 km square centred on La Jolla, almost 200 species of native flora (not including all the introductions, which are many) have been recorded. And in 2014 the *Checklist of the Vascular Plants in San Diego County* documented 2,447 species (including hybrids).

Plants are many-celled organisms that make their own food by harnessing the sun's energy. Sunlight is captured by a green pigment, chlorophyll, contained within structures in plant cells called chloroplasts. Unlike animal cells, the walls of plant cells are made from cellulose and the overall structure is protected by a cuticle. They have specialised organs for reproduction. As we'll see, the gulf between the number of species described and how many we estimate are *out there* is often colossal, with estimates judged as *reasonably poor* to *very poor*. However, we do have a pretty good handle on how many species of plant there are. There are currently almost 400,000. Most of them (90%) are flowering plants (or angiosperms), with 6% being mosses, almost 4% being ferns and horsetails, with the conifers coming in at 0.3%. Our estimate of plant species richness could be described as *very good*, largely because, for the most part, they are large, they stay put, and there has always been a huge and long-standing interest in them as sources of food, building materials and medicines (see Chapter 5). The botanists reckon there are no more than 60,000 species to find and/or describe, even though new plant species seem to be discovered on a fairly regular basis. And there are still surprises – not least a new 105-tonne species of tree, and a 1.5-metre tall carnivorous plant, both described in 2019. To the nearest approximation every plant is a flowering seed plant, an angiosperm.

Now to the seaweeds – placed here because you might expect to see them here with the plants. On the beach at Bird Rock there are well over one hundred different species of seaweed with at least three of those species being invaders. But seaweeds are not plants – they are not, it turns out, even closely related to them.

Seaweeds have considerably more complicated life (his)stories than plants. They have a holdfast, an attachment for securing the seaweed to a hard surface: rocks, animals, even other seaweeds. These structures are often mistaken for roots, but the holdfast is for attachment only and cannot take up nutrients. Like plants, seaweeds contain chlorophyll but the three different colour varieties that define the main groupings are based on other pigments. The green and red seaweeds are more closely related to plants than brown seaweeds are.

While they live submerged in water, seaweeds are not normally found at depths greater than 100 m, below which sunlight does not penetrate enough to drive photosynthesis. When I've been at Bird Rock in the spring the red seaweeds (Fig. 5.2b) dominate the shore both in terms of cover and species richness. Near the top of the shore the brown seaweeds, namely channelled wrack, *Pelvetia* (Fig. 5.2c), and string-like *Scytosiphon* are evident, with a number of other brown species below them. Those browns make up about 20% of the seaweed species present. Also, just below the channelled wrack are thick mats of bright green *Enteromorpha* and sea lettuce *Ulva* (Fig. 5.2d), the green seaweeds generally being one in every ten species encountered. The Californian coast has a diverse seaweed flora with 769 recorded species. This figure is based on the *California Seaweed eFlora*, a web-based update (almost ready to be released at the time of writing) of the 1976 classic *Marine Algae of California* by Abbott and Hollenberg. Worldwide there are about 6,200 described species of seaweed. Globally, as locally at Bird Rock, the red seaweeds are the most species rich (66%), followed by the browns (24%) and then the greens (10%). Many species are edible. They are used as fertilisers or sources of sugar. They can be microscopic or can form extensive kelp forests, creating entire ecosystems, like those that are found in the shallow waters off SoCal. The giant kelp *Macrocystis* can be up to 33 m in length.

3) Fungi: Mushrooms, moulds and yeasts – The Fungi

Not actually on, but on the way down to Bird Rock beach, amongst some rough grass, I noticed what looked to me like some mottlegill mushrooms, each with a bell-shaped brown-coloured cap about 1 cm across (Fig. 5.3a). These mushrooms are probably the most prominent and well-known members of a group that is turning out to be extremely species rich, greatly increasing the diversity of microbes, the Fungi. The mushroom itself is actually just a fruiting body. The actual organism is living sort of dispersed in the soil, which is why fungi are lumped with microbes. I haven't seen many mushrooms in SoCal but then I've not been there very much after periods of heavy rain, which mushrooms love. On the beach proper there are likely to be relatively few marine fungal species. They are the arguably less well-known and microscopic moulds and yeasts (Fig. 5.3b). That said, I have occasionally seen mushrooms grow on other beaches. The prediction of poor microfungal diversity on Bird Rock is based on studies on a similar beach a little further south. There, 52 species in total have been recorded, but only 10 of those would be classed as marine. A recent inventory of intertidal fungi in China recorded over 6,013 species, although it wasn't clear (to me at least) which ones are exclusively marine. So maybe it's still too early in terms of marine fungi to say anything.

There do seem to be more land-living fungi than marine. Of an estimated 1.5–5 million species, only 98,998 have been described so far. Of the described species, just over 1,100 are marine (although total marine is estimated to be an order of magnitude greater). Other authorities give estimates of 2.2–3.8 million species of fungi, still more than five times the number of plant species. However, the bottom line is that because the group is comparatively poorly known, making good predictions is difficult. Many people make the mistake of lumping fungi with plants. Ironically, in evolutionary terms they seem more closely related to animals than plants, and are so different from both that they are rightly placed out on their own, as we shall see later. They digest their food externally using enzymes they excrete into the environment. Then they absorb broken-down organic stuff directly into their cells. Their cells are mostly made of chitin, the same stuff that makes up the outside skeleton of insects and lobsters. Many fungi cause diseases but many more form constructive, intimate associations with plants. Such fungi are near essential for plants to grow in nutrient-poor soil. Moulds and yeasts are used extensively in the production of cheese, beer, wine and soy sauce. Antibiotics, such as the mould penicillin, are now a key part of healthcare. Penicillin mould also develops flavour in Stilton, Danish blue and Gorgonzola cheeses. Arguably, the largest creature alive on Earth today is a fungus called *Armillaria*. There is a living clonal growth of this fungus in the US state of Washington that covers an area of 600 hectares, and one in Michigan weighs more than 100 tonnes. That's about the same mass as a blue whale. Both of these fungi are estimated to be more than a thousand years old. Of course, most fungi are much smaller, and often we are unaware of them until we see their fruiting bodies or until they expensively turn the timber in our house to mush with those nightmare words *dry rot*.

4) Mollusca: Shell life

Walking onto the foreshore at Bird Rock there are a number of things which immediately strike you. The huge amount of seaweed blanketing the shore, particularly in the summer. But so too does the sheer number of shelled animals on the surface of the rocks, in depressions, in the crevices, and in the tide pools. There are lots of different shelled snails – limpets, turban and top shells, periwinkles (Fig. 5.4a) and dog whelks – and they are in abundance. Clumps of Californian mussels (Fig. 5.4b) are obvious on the surf-exposed rocks. But there is also a rich variety of relatives of these animals which repay careful, patient searching. Chitons (Fig. 5.4c), or sea cradles, as they are sometimes known, blend in with the rocky depressions in which they are often found. There are numerous snails that have lost (over time, not that they've misplaced it or been mugged) or reduced and internalised their shells – the sea hares (Fig. 5.4d), the sea lemon and many more beautifully coloured, elegant sea slugs. And if you are fortunate you may even find the stunning and captivating two-spotted octopus *Octopus bimaculoides* (Fig. 5.4e) in a crevice or beneath a ledge in a tide pool at the bottom of the beach. In the report on the San Diego–La Jolla Ecological Reserve mentioned earlier, the brief inventory of shore invertebrates is dominated by molluscs. They are so obvious. And so on this beach the shelled animals (and their relatives) seem as species rich as the arthropods, although the arthropods probably still outstrip them if all the microcrustaceans are taken into account.

These shelled animals belong to the phylum Mollusca (from the Latin *molluscus*, 'soft', referring to their bodies not their shells). This is a group comprising snails, slugs, chitons, clams and cuttlefish. All molluscs have a muscular foot attached to a soft body containing the internal organs. There is a fold of tissue (the mantle) that drapes over the main body mass and is responsible for producing a hard shell made predominantly from chalk, calcium carbonate. Most species of mollusc are marine but they are also very successful both in fresh waters and on land. It's difficult to get a good estimate of their species richness. A recent study recognised 46,000 valid live marine species and suggested that another 104,000 remain to be described. Large specimens, primarily bivalves (clams, mussels and the like) and cephalopods (squid, octopus, cuttlefish), can be measured in metres, although some of the smallest require a microscope to see them, and there is everything in between. Squid, oysters, scallops, cockles, periwinkles and land snails are highly prized as food, and some species, and their products, are considered as objects of natural beauty (e.g. mother-of-pearl and pearls themselves). It's not all positive though. Some species are pests. Many snails act as intermediates for the infection of other animals, including humans, by parasites (e.g. 290.8 million people required preventative treatment for bilharzia in 2018). And the shipworm *Teredo* is legendary as it effectively destroys untreated submerged wooden structures such as piers and boats.

5) Chordata: Animals with backbones...mostly

The most common and species-rich backboned animal (vertebrate) around Bird Rock and worldwide is, arguably, the fish. There are many species that live between the tides. The woolly sculpin (Fig. 5.5a) seemed pretty abundant each time I've been there. And there's the small and unscaled rockpool blenny swimming in tide pools. It is most noticeable by its overenthusiastic, overlong eyelashes. And beneath rocks we have the tadpole-shaped clingfish, much larger than the blenny and fitted with a disk-like sucker on its underside with which it clings limpet-like to rocks. I have seen at least ten species at Bird Rock which specialise in this specialist intertidal life, but there are also more fleeting visitors to the shore – particularly during the high-tide period. All of these fish species are teleosts, or bony fish – as far as I'm aware the only cartilaginous fish frequently encountered in this region is the leopard shark. According to the *Guide to the Coastal Marine Fishes of California* there are over 550 marine fish in Californian waters, with 439 of those found in shallow waters.

The bird life of Bird Rock, perhaps not too surprisingly, is quite spectacular. Dark-plumaged, endangered brown pelicans (Fig. 5.5b) and other birds can be found perched on Bird Rock itself, the focus of a plethora of amateur and professional photographers. When I was last there in February 2020, without really looking I noticed cormorants, finches, grebes, gulls, hummingbirds, sandpipers and of course the brown pelican. The eBird website records at least 81 bird species at Bird Rock. Currently, around 700 species of bird can be seen in California, and many of those in coastal locations. Sadly, the number of species is not as great as it was, as the coast is so heavily impacted by development and recreation activities.

Just a little further along the coast at La Jolla Cove there are the highly sociable Californian sea lions (Fig. 5.5c) (they have ear flaps, and can walk and climb), huddled together on the rocks next to the caves. Not too far away are the more solitary harbour seals (no ear flaps and flop about on their stomachs). Both sea mammals attract a large number of tourists when in residence. A number of snorkellers and divers visit the red kelp beds nearby in La Jolla's Marine Conservation Area. They call it 'Turtle Town' because green sea turtles (Fig. 5.5d) go there to feed on the kelp. And giant whales are not exactly uncommon off the coast at some times of the year.

Living vertebrates are made up from about seven different, though related, family lines, six of which we've mentioned. Three of them are obligate water-dwellers and have come up with a fish-shaped body form independently from one another: the jawless fish, such as lampreys; the cartilaginous fish, the sharks and the rays; and the bony or ray-finned fish, the pipefish, stickleback, trout, wrasse, perch, flounder and the like. The most species-rich group is the perch family. To the nearest approximation every bony fish is a perch and so too is every fish. In fact, about half of all vertebrate species are fish. There are an estimated 50,000 vertebrate species with 45,000 already described. It is believed that, as with the plants, this estimate is pretty good.

There are three groups which are, for the most part, not aquatic, although they have some members that have returned to the sea: the reptiles, the birds and the mammals. For the group we belong to, the mammals, most of the groupings it contains are simply not very diverse, with few exceptions. To the nearest approximation every mammal is a rodent. 'Just bats and rats' I was told by a doctoral student during her *viva vocal* examination when I questioned her about the huge mammal diversity in her Central American country. Straddling a watery and a land existence are the amphibians. Their aquatic young stages have a typical fish-form, whereas the land-living adults more closely resemble the reptiles and mammals in their body form, although, unlike these two groups, adult amphibians are poorly waterproofed. There are of course some mammal groups that have evolved a fish-like form: the whales, dugongs and the manatees. They are obligate water-dwellers, even though they must return to the surface to breathe.

All of these vertebrate groups belong to the Chordata, though not all chordates are vertebrates. There are two smaller chordate groups that do not have a backbone. When included in the species tally, they bring the number of described chordate species up to 64,788, with an estimated species richness of 80,500. The first of these non-vertebrate chordate groups is the Urochordata, the sea squirts (Fig. 5.5e) and salps. The group encompasses solitary forms like the brown-stalked sea squirt *Styla*, or colonial forms such as the beautiful blue star sea squirt, but also the huge, jellied colonial salps. And secondly there are the Lancelets, the Cephalochordata. They basically look a little like fish...but without jaws, without eyes, without fins...well, nothing like a fish really. But they do illustrate in a very easy-to-see way all the characteristics that make a chordate a chordate. So, although there are very few species, they are famous for being used in zoology teaching labs all over the world because examining one 10-centimetre-long individual it is easy to see its blocked, segmented body muscles, its dorsal hollow nerve cord, its gills, its post-anal tail (just think, nearly all other animals have their anus at the very end of their bodies – don't hold that thought) and the presence of a cartilaginous rod, the notochord, running the length of the body, nearly all of the features that make a chordate a chordate.

6) Protozoa or Protista?

When you say you're a marine biologist, most people's eyes grow wide as they say, 'Wow – do you study sharks/whales/dolphins?' When greeted with, 'Eh, no, I study beachfleas,' there is usually an awkward pause and a drifting away of sorts. On one occasion (in 1994) I was at Bird Rock looking for mites living in the groove on the underside of a beachflea (Fig. 5.1b), mites similar to my *namesake* (*T. spiceri*) (Fig. 5.1f) that I told you of earlier. I found some but they were completely outnumbered by hundreds of single-celled organisms – ciliates, both attached to the limbs underneath the beachfleas' body (Fig. 5.6a) but also crawling (Fig. 5.6b) amongst swathes of miniature filamentous green algae and what I took to be fungi. (Ciliates are a diverse group but all of them have bodies covered by, or bearing patches of, minute hair-like cilia which beat and propel water through the organism (attached ciliates) or the organism through water (crawling or swimming ciliates).) On this occasion my response to the enquirers was, 'I'm studying these little beachfleas and each one has its own little ecosystem on its belly.' This time there was genuine interest, although admittedly they did still take a step back. You need a good magnification device to see these ciliates, these single-celled protozoans, and so the enquirers had to take my word on what was there. Truth be told these organisms were almost everywhere you looked on the beach, if only you had a good magnifying device. In fact, they tend to be everywhere you look wherever you look – and there are always lots of them.

Up until recently, protozoans, as the name suggests, were thought of as primitive animals. However, it has long been known that some have animal-like and others plant-like characteristics. The term protozoan, while considered obsolete by some, is now restricted to the animal-like forms, forms that ingest food. To cover the variety of so-called single-cell life, the Kingdom Protista was erected (see p. 53), but in reality it was not a natural group. It was a catch-all bag for things that did not fit anywhere else. So protists consisted of the protozoans, one-celled algae (so the seaweeds are sometimes brought to join this group) and the slime moulds. The whole arrangement, as we'll see, is unsatisfactory, but it'll do for now as we try to enumerate the groups involved.

Waters just north of La Jolla, in the San Pedro Channel between Santa Catalina Island and Los Angeles, contained hundreds, sometimes thousands, of protist OTUs (operational taxonomic units), with different OTUs found in shallow compared with deep water. Protists are also species rich in the waters local to the Marine Laboratory in Plymouth, UK (almost 500 species) and off the coast of China (about 3,000 species). Although there are protists floating in the water it has been suggested recently that those living on the beaches and on the bottom of shallow seas probably contain most protist diversity.

And then there are the protozoans that live within the organisms found on the beach. For example, investigation of ten species of limpet and turban snails, present on Southern Californian shores just north of La Jolla, found they were infected by six ciliate protozoans. There may well have been more protozoans present but the work, carried out by a student working to get their doctorate from UCLA, concentrated just on one group, the ciliates, and was mainly concerned with investigating the species specificity of this group using cross-infection experiments. In the waters around Plymouth, just over one third of the protozoans recorded were retrieved from within organisms. The 'species richness' of protozoans in the environment, and in the other organisms, of and around Bird Rock beach is likely to be numbered in the thousands.

In terms of how they're constructed and how they function, the Protozoans can be considered the most diverse of all eukaryote forms of life (we'll come to this shortly). While only around 40,000 species of protozoans have been described, the best estimate is about 200,000. It is generally recognised that these are pretty poor estimates, but they substantially bump up the species of microbes generally. Protozoans are found everywhere there is water. In salt and fresh waters, living on the bottom or as plankton (free-floating inhabitants of the sea or lakes), in damp soils and leaf litter, and inside every other living thing, including you and me.

7) Nematoda: The roundworm that's the fly in the ointment?

The first free-living marine roundworms were collected from SoCal in 1915 by the famous (amongst marine scientists at least) Th. Mortensen Pacific Expedition. And yet a century later we still have a very poor picture of roundworm diversity here, and indeed on most coasts. Certainly, only a relatively small number of nematode roundworms have been formally recorded from beaches around La Jolla. Many are found on animals and seaweeds, particularly within holdfasts, the attachment structures of seaweeds such as kelp. That said, I've seen a number of different types when I've been high up on the beach at Bird Rock. Not that I could put names to them. They were in small pockets of sediment, and beneath the flotsam, jetsam and weed cast up at the high-water mark, and on and in many beachfleas that live at the top of the shore. Their thrashing back and forward is unmistakable, and easily recognisable as long as you can see the animal.

Nematode roundworms (or threadworms or hairworms) look, on the surface, a bit like earthworms but with no segments.

Inside they show none of the repetition of structure that we see in earthworm segments. When very small, they look just like individual threads (Fig. 5.7a), hence their name (the Greek *nema* means 'thread'). They have a tough outer skeleton made of chiton, one of the key components of arthropod exoskeletons, and they have a simple inner muscle arrangement which is why they thrash about a lot.

Now, nematodes are notoriously difficult to identify. It is interesting that someone who did know how to identify them, Dr Wolfgang Weiser, made an extensive collection at one location, in the intertidal zone, right in front of the Marine Laboratory on the Hoe at Plymouth. He found almost seventy species in a very small area.

So, given the difficulties of working with a group where peering at their genitalia down a high-power microscope is an essential part of identification (although genetic barcoding is now being used to great effect for ID), it should come as little surprise that while less than 25,000 have been described, the currently accepted species richness is estimated at 500,000. That's why the nematodes look as if they are in the wrong rank place in Figure 4. Interestingly some other estimates of species richness have varied between one and, yes you are reading it correctly, one hundred million species. Certainly, the accepted estimate is poor. But the idea that a section of deep-sea floor has 100 million species, and so is more diverse (in terms of numbers of species) than a rainforest, is now waning...a little.

Nematodes are extremely prolific. Some species are able to produce 100,000 fertilised eggs every day. It has been estimated that there are 100,000,000,000,000,000,000 individual roundworms on Earth, and this doesn't take into account species that are parasites of plants and other animals. Most free-living species are quite small (millimetres or less) but this is not so for all the parasitic species where centimetres, and occasionally metres, is the measurement unit. The largest known roundworm was found in the placenta of a female sperm whale. It was nine metres long. Probably the best-known roundworm is the soil-living *Caenorhabitis elegans*. As part of the Human Genome Project the genome of this *representative invertebrate* (animal without a back-bone) was sequenced and published. With a size of 80 Mb, it is one of the smallest animal genomes sequenced to date.

When it comes to roundworms known for bad reasons, the possibilities are endless. A good number of different species parasitise and damage key crops such as cereals, cotton, citrus and other fruits, legumes, potatoes, sugar beet, trees and so on. While there is a similarly large list for animal parasites, it is with humans and their domesticated animals that the effects are felt, sometimes literally. There are a number of tropical, blood-sucking intestinal hookworms. There is *Ascaris*, which lives in the small intestine where it 'interferes' with its host's ability to take up food. And *Wuchereria*, which blocks the lymph channels resulting in the disease elephantiasis. Many parasites require a carrier animal to infect the person, or the pig, or the horse. For instance, the roundworm *Onchocerca* is transmitted to humans by blackflies, and infection results in tropical river disease, and in some cases *river-blindness*. There are an estimated 24,000 nematode parasites living in individuals of a taxonomic group we know quite a lot about – the vertebrates. Even here most of those species are still to be described.

8) Bacteria and Archaea: Microbial life

Bird Rock is literally packed with what is likely to be hundreds, if not thousands, of microbe 'species' (or OTUs). Microbes are cellular organisms that are *nearly always* too small to be seen with the naked eye. And there are lots of different types. Attempts have been made to characterise the microbial communities in the waters adjacent to the pier at Scripps Institution of Oceanography and Mission Bay (remember the volleyball). Breitbart and colleagues found that only 24–33% of the genetic material detected could be linked to existing records of microbial life. What was also interesting was there were some fundamental differences in the composition of the samples even though they were taken less than 7 km apart. Analysing genetic material extracted from sand samples, collected from 49 sandy beaches spanning the Californian coast (and including some very close to Bird Rock), unearthed an incredible amount of previously undetected diversity. Each beach was characterised as having between 639 and 2,750 unique microbe OTUs, giving the total number of OTUs in the sands of the Californian coast as 48,543, distributed between 42 very different major groupings which the investigators referred to as *phyla* (see p. 59). Similar results have been recorded for other parts of the US, including Florida and Hawaii, as well as from the coast of China. And, of course, sand is just one of many different *environments* at Bird Rock where we will likely find microbes. That diversity could quickly become mind-boggling if we were to find that different types of microbes live on rocks, in crevices, in tide pools, and on and within the huge variety of animals, plants and seaweed on the shore and in the coastal fringe. While the presence of microbes associated with sewage can give an area a bad name, many of the marine microbes play vital roles in the local and global cycling of nutrients, matter and energy in our ocean.

Despite all this talk of megadiversity there are currently only about 13,000 described species of microbes, which belong to either the Bacteria (Fig. 5.8a) or the Archaea. Together the two groupings constitute one of two fundamental patterns of life, and are termed prokaryotes (details of what makes a prokaryote a prokaryote follow shortly). However, as we'll see a bit later, the relationship between the two groups is much more distant than we had previously thought or even could have imagined.

The largest microbe group by far is the bacteria. They are ubiquitous – they are literally found everywhere. Bacteria are nearly all very small, microscopic critters between 1 and 5 millionths of a metre in length. The largest species is 0.75 mm in diameter. Discovered in 1997, it is a marine bacterium, *Thiomargarita namibiensis*, the sulphur pearl of Namibia. It gets its name from the fact that it uses sulphur in its life processes and inhabits coastal mud off Namibia, south-west Africa. About half of the species known are capable of moving in a particular direction, and at speeds up to a hundred times their body length per second. They are extremely numerous. I once read that the total number of bacteria in your mouth is greater than the number of people who have ever lived. One tenth of your dry mass is made up of bacteria. It has been estimated that the world contains 5,000,000,000,000,000,000,000,000,000,000 individual bacteria and this number is growing fast. Bacteria can be free-living, inhabiting the surface of other living things, as well as making their home within animals, plants and other bacteria. In terms of what they do for us, it is difficult to say. This is mainly because they seem to be important for so many different reasons: essential to health of digestive systems, soil maintenance in agriculture and forestry, and even the existence of the air we breathe. Truly, the bacteria are the hardiest of living beings. Some can survive extreme temperatures including boiling hot springs.

As previously mentioned, until recently bacteria were lumped together with a much less species-rich group of what at the time were believed to be very *primitive* microbes, the Archaea. Archaea are generally capable of surviving extreme conditions and so are often found in *unusual* places such as hot springs or super-salty (hypersaline) waters.

So, if the Bacteria and Archaea are so numerous, why are they only at number eight in our list, and not at number one? Well, one major problem with investigating such mini-life is, as we have discussed, how to recognise a species. In fact, for many microbiologists Cowan's long-standing view still holds, that 'the microbial species does not exist…it is one of the great myths of microbiology.' We said earlier in the chapter that morphological species were by far the easiest to use in most circumstances. But this is certainly not true for such tiny critters where in one sense there is very little to see. Bacteria are very *basic* to look at, often being superficially sphere- or rod- or spiral-shaped. So, biologists initially resorted to chemical tests to supplement the differences in what particular bacteria look like. Even then, you need to be able to grow the bacteria in the laboratory to test it, and most bacteria we simply have not grown, or cannot get to grow, under laboratory conditions. More recently, differences in the composition of genetic molecules such as DNA have been used to discriminate between different 'species' or OTUs, and this technique is proving extremely powerful and useful, although not always as precise as we would like (or as some people imagine).

As mentioned, there has been a shift in thinking about how biodiversity is distributed, with microbial diversity (in terms of 'species' richness) exceeding that of the larger organisms. A recent claim of Earth being home to a trillion microbial species really would, if true, be an even greater game changer. Such grand claims are not new, with estimates of prokaryote species numbers rising, through the 1990s, to millions and even billions. However, a census of prokaryote species published by Rudolf Amann and Ramon Rosselló-Móra in 2016 and entitled 'After all, only millions?' seemed to indicate that the figure might actually settle out in the lower millions rather than the billions, as there seems to be greater similarity in microbe types than indicated by earlier identification techniques. As an onlooker one feels that this debate is going to run for a bit yet, and is still far from settled. The Bacteria and Archaea look as if they will be the most 'species'-rich group on Earth. But by how much will they exceed the arthropods?

Remaining animal groupings

All of the remaining five groups on Figure 4 together with a further 45-odd phyla (not presented on the graph) are an eclectic mix of over 50 fundamentally different body designs. Together they total an estimated quarter of a million species, although only 115,000 have been described to date. However, just because these species didn't make the top four animal groups doesn't make them uninteresting or unimportant. Taking a different measure of biodiversity may well have produced a different ranking order. Indeed, some of the most noticeable, and noteworthy, animals found at Bird Rock do not belong to the four species-rich animal groupings previously set out. For example, the underside of many of the rocks are carpeted with the white outer shells of numerous filter-feeding worms. They belong to the segmented true worms, the Annelida (Fig. 5.9a). The 16,763 described species exhibit a huge diversity of forms and habits, and include key species that are important in that they aerate sediments. And then there's the 6,000 species of sponges, the Porifera. They have retained the most primitive features of all animal forms, but with internal support provided by a bewildering diversity of often incredibly beautiful and intricate little structures called spicules, star-shaped, pen-shaped, and even anchor-shaped. There are creatures that look like delicate little plants, the sea firs, relatives of the sea anemones, and sea anemones themselves like the abundant *Anthopleura* (Fig. 5.9b), and jellyfish and corals. And how could we miss the knobby seastar *Pisaster* (Fig. 5.9c) belonging to a group of spiny-skinned animals, the echinoderms – what I like to refer to as tiny animals living in huge bodies. The grouping has only about 7,300 described species, with estimated species richness ranging from 10,000 to 13,000. That said, they show incredible diversity, ranging from the flower form of the ancient sea lily, through to the sea urchin and starfish, ending up with the peculiar, elongated, sausage-like sea cucumber. In fact, if we were considering the biodiversity of the deep sea on its own, the echinoderms could conceivably be up there near the top.

Viruses: All the world's a phage...or nearly

Last on is a 'life form' that for many years we've not known what to do with – the virus (Fig. 5.10a).

Maria Breitbart and her colleagues investigated the marine viruses present in just 200 litres of surface water collected from the end of the pier at Scripps Institution of Oceanography, just north of Bird Rock, and the same volume from water close to Mission Beach. Each 200-litre sample contained 10^{12} virus particles. They estimated that between the samples there were 374–7,114 viral types. So, whatever a virus is, there are a lot of different types and a lot of individuals of those different types, even in a relatively small amount of seawater, let alone the ocean.

All of the life forms described up until now are either cells or are made up of cells. They are living in the sense that they have all of the machinery required to replicate and carry on the family line. The virus does not quite fit the 'living' category. They've been described as being 'at the edge of life'. Each virus is composed of, quite literally, bits of genetic code (DNA or RNA) wrapped up in a protein coat. They seem to be more closely related to the cells in which they reproduce than to one another. And it is this returning to already living tissue, and being dependent upon its complex structures and chemicals to enable them to replicate, that marks viruses off as something quite different.

Certainly, in the last 25 years the uncovering of a huge diversity of marine viruses has transformed our understanding of how the ocean works. We've just discovered it is interactions involving the viruses that give seawater its distinctive smell. And every living thing at Bird Rock is going to have its own complement of specific viruses. It is thought that viruses are the most common biological 'entity' in our ocean. Marine viruses contain more carbon than 75 million blue whales. If you joined the viruses end to end, they would stretch out 10 million light years – that's past the nearest 30 galaxies.

At present, I don't think we have a good handle on how many viruses are present at Bird Rock but, in truth, this mirrors our current global inventory. Approximately 5,000 viruses have been formally described, but the estimated species number is a hundred times that. And even that is too conservative an estimate for some. Most viruses are associated with bacteria and we refer to them as phages. The fact that viruses seem to be more closely related to the species they are found in than to one another means that as a minimum estimate there could be as many viruses as there are species of living things. This assumes of course that each species has produced only one virus. It also assumes that species don't share the same virus. In 2019 Carlson and colleagues found that there were *only* 40,000 virus types in vertebrates, reducing previous estimates by two orders of magnitude.

Where there is life there are viruses. They are a major cause of mortality it is true, but it is increasingly recognised that they play key roles in how ecosystems work. I don't think it would be possible to find any (reasonable?) human being on our planet at this time of writing (September 2020) who doubted the possible far-reaching effects of viruses, given we are currently in the middle of a COVID-19 pandemic.

Given how little we know about species diversity of viruses, we are quite knowledgeable about their genetic diversity. Viruses are regarded as the greatest repository of genetic diversity on the planet. The genome (all of the DNA taken together) of many hundreds of viruses has been mapped. This is principally due to the small size of their genomes. Hepatitis B is one of the smallest genomes at 3,200 base pairs (bps) long, and smallpox one of the largest at 186,100 bps. Compare this with the genome of the mouse (2.9 billion bps) or even a humble fruitfly (139.5 million bps).

New species

We are still a long way from having a complete catalogue of all the living beings with which we share this planet. Species continue to be described but not in any systematic way. The groups that receive attention do so as a result of individual scientists and amateur enthusiasts choosing to work on their favourites. So most new descriptions are of insects and spiders, groups which command a wide interest from professional and amateur alike. The main exception to this is when a group of organisms is of medical or economic importance. By and large, until recently, the majority of new species described were relatively common (not rare) and comparatively large (not microscopic) life forms that were collected mainly from temperate (not tropical) regions. However, over the last quarter-century, the number of new species of Bacteria and Archaea has increased to 600–700 new descriptions each year, and that number is still rising. About 1.5 new species are described every hour (13,500 each year). But not all new species belong to obscure groups. New fish species are being described at a rate of just over ten a month. For birds it's one species every two months. And nearly 2,000 new species of plants, and an equivalent number of fungi, were found in the period 2016–17. Even within the mammals, about one new species is described every three years, with a relatively high proportion being aquatic forms such as whales and dolphins. Description of a new species is not uncommon even in areas where the wildlife is thought to be well known. For example, in 2020 a common and conspicuous worm, which has long been known to live in the upper part of the La Jolla submarine canyon, was finally formally described and named *Chaetopterus dewysee*. This in spite of the fact that it has been used at Scripps Institution of Oceanography for a number of years now as a model species for studying biological light production.

But still, the rate of description overall is comparatively low. Indeed, we may be facing a situation where some groups of species are becoming extinct faster than they can be described. Putting together catalogues of species is not just hampered by the rate of species description but also by the fact that sometimes different species are given the same names, and sometimes the same species is given different names in different locations. Although these two activities are bound to cause confusion, we still do not know the extent to which each is a significant problem. It has been shown that one fifth of insect species have more than one scientific name. And it's not just at the level of species description that we see change and rethink. As we shall see next, the higher classification of living things is far from settled and is still the subject of much debate.

Planting and growing the 'tree of life'

The process of classifying species (describing them and putting them into a larger framework) and the product of that process is referred to as the study of systematics or, to use an older word, taxonomy. Thus, taxonomy is one of the mainstays of biodiversity research. There have been numerous attempts throughout human history to place all the elements of the living world into meaningful, discrete and useful categories (Figs. 6–11).

Figure 6 Aristotle's 'great chain of being'. Figure is Ramon Lull's *Ladder of Ascent and Descent of the Mind* (1305). (Courtesy of Wikimedia Commons.)

The great chain of being

Plato (428/7–348/7 BCE) and Aristotle (384–322 BCE) came up with, what by the Middle Ages had developed into, 'the great chain of being' (Fig. 6). We talk today of holistic and inclusive approaches to a subject but Aristotle's scheme, and its derivatives, remind us that this is far from a fad. His classification scheme included what he believed to be all life, natural and supernatural. In Aristotle's *Study of Animals*, he says, 'Nature proceeds little by little from things lifeless to animal life...next after lifeless things in the upward scale comes the plant...there is observed in plants a continuous scale of ascent towards the animal... And so throughout the entire animal scale there is a graduated differentiation in amount of vitality and in capacity for motion.' One of the most interesting points about 'the great chain' is that it is not just a set of helpful box labels, where an entry can be easily retrieved when required. It is in fact an antiquarian world view, which lent itself nicely to Christian thought in the Middle Ages, carefully constructed to show the interrelations and dependence of its elements on one another.

Linnaeus's hierarchical classification

Dubbed the father of modern classification, Carolus Linnaeus came up with a hierarchical organisation of life that, to a large extent, still has considerable influence on how we group living things today. Linnaeus, reflecting a world view common in his time, believed that all species, or kinds, had been created separately by God. Thus, there were no relationships to reflect. So to a large extent how one grouped living things was a matter of personal preference. It is a bit like working in a DIY shop and being given numerous types of nuts, bolts, washers and so on to put on display. It is a good idea to arrange them in some sort of order, according to thread size, geometry or function, but there is no one way to do it. In the same way, organisms were placed into human-made categories based purely on how they looked. It was Linnaeus who introduced the terms kingdom, class, order, family, genus, species. He divided living forms between two kingdoms, the Vegabilia and the Animalia. Thus, the two-spotted octopus, *Octopus bimaculoides*, which I was fortunate (and tickled) to encounter in a pool on the low shore at Bird Rock, belongs to the genus *Octopus*, which is placed in the family Octipodidae. The Octipodidae is one of many families that belong to the order Octopoda, which contains many different groups of octopods. The Octopoda

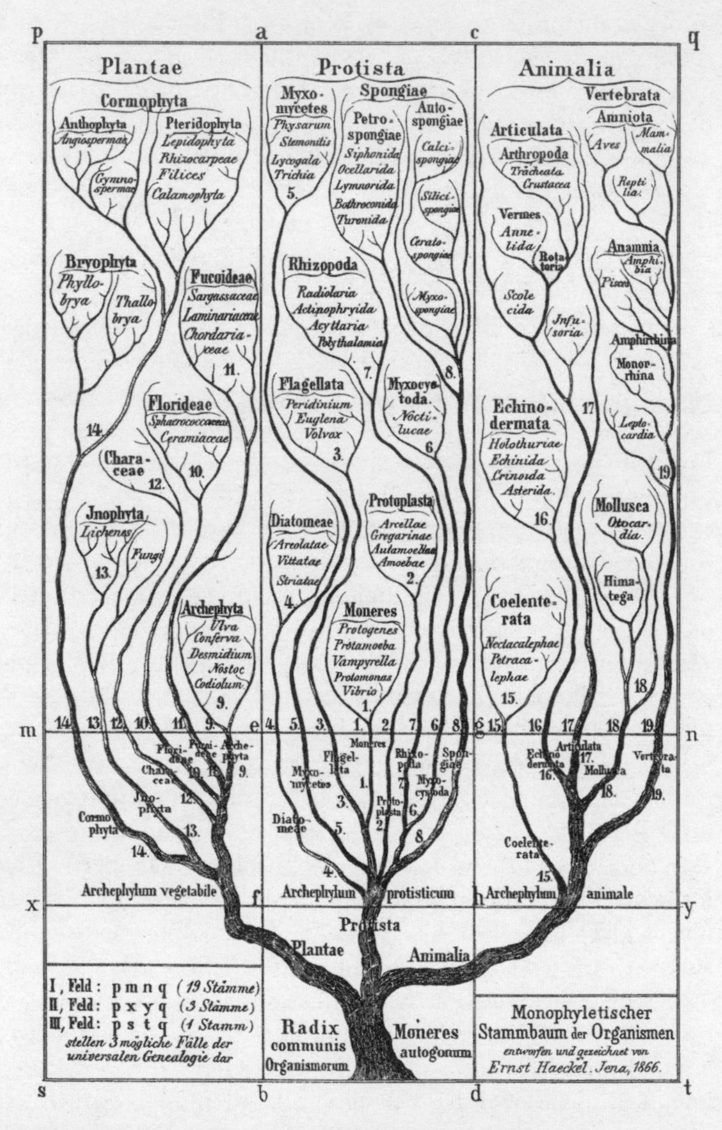

Figure 7 Haeckel's tree. (Courtesy of Wikimedia Commons.)

belong to the class Cephalopoda, which is a large grouping that includes octopus, cuttlefish, squid and the shelled *Nautilus*. The Cephalopoda sit within the phylum Mollusca, which includes all snails, slugs, clams and chitons, and the Mollusca belongs to the kingdom Animalia – all of the animals.

Influence of evolutionary ideas

The publication of Darwin's *The Origin of Species* in 1859 heralded a new way of thinking about classification. It was a bit like finding out that the nuts and bolts in our DIY shop were in some way related to one another, and thus how they were arranged could actually reflect those relationships. Biological classification was no longer just a case of assigning life forms a label, using a pragmatic system where easy access to material was the primary aim. Species were related to one another. They had common ancestors. So the emphasis shifted in the nineteenth century and the century that followed to elucidating evolutionary trees. They were diagrams, sometimes literally drawn out by authors as trees, where species were placed on branches in a way that illustrated exactly how they were related to one another (Fig. 7).

The kingdom, phylum etc. groupings of Linnaeus were retained and extended by the 'German Darwin', Ernst Heinrich Haeckel. He introduced a third kingdom, the Protista, for microscopic organisms. But now these designations had to, as much as possible, reflect evolutionary affinities. What began to emerge was the idea of life being like a tree, sending out more and more branches as it grew. Right up until the present day there still exists a tension between those who are interested in classificatory schemes which reflect evolutionary relationships (cladistics) and those that *merely* want to make it easy to identify an unknown organism. Arguably, one is not intrinsically any better than the other. It all depends on what you want the scheme for.

If you want to put a name to a face then a carefully worked out evolutionary scheme is perhaps a bit of overkill and may actually make what should be a simple identification process very complicated.

Chatton's two-domain idea

In a now famous 1962 paper, 'The concept of a Bacterium', Roger Stanier and CB van Niel suggested that life could be divided into two major groupings based on two fundamentally different types of cell organisation. Recognition of this division was fuelled by the development of new techniques following the Second World War and a rapidly advancing understanding of the way different cells were organised. Stanier and van Niel credited Edouard Chatton with the original idea – he had first recognised this division in 1937 but had been overlooked. Chatton proposed that life could be split into just two domains, the prokaryotes and the eukaryotes. All bacteria were assigned to the prokaryotes. Everything else that is alive, the animals, the plants, the fungi, the one-celled microscopic organisms, were assigned to the second main division, the eukaryotes.

Prokaryote cells are smaller (0.2 to 10 millionths of a metre) than eukaryote cells (10 to 100 millionths of a metre), and do not have smaller organised structures inside them. Prokaryotes do not have mitochondria, which are the powerhouses of the eukaryote cell, or plastids, which are structures that plant-like organisms have where photosynthesis takes place. There are many other differences but perhaps two of the main ones are: 1) in prokaryotes the genetic material that carries the plan for building new individuals is a single strand of DNA, not contained in a cell nucleus, whereas in eukaryotes the genetic material is a combination of DNA, RNA and protein arranged in 2–600 chromosomes, all of which are enclosed in a membrane

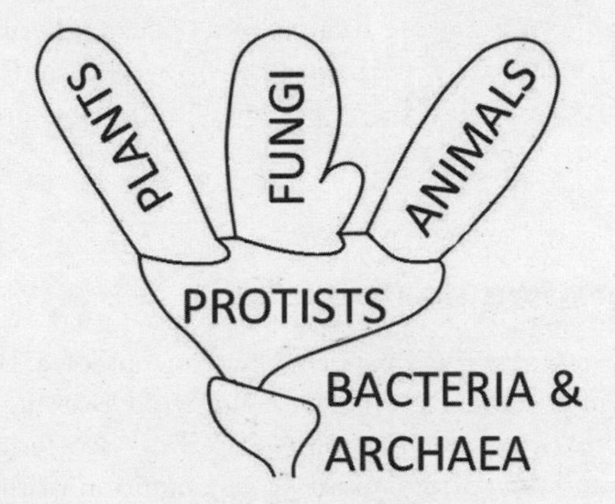

Figure 8 Whittaker's five-kingdom approach.

that constitutes a structure we call a nucleus. The word eukaryote means 'true nucleus'. 2) In terms of how they get their food and make energy available to do things (metabolism), the prokaryotes, and so the bacteria, are probably the group that shows the greatest diversity.

Whittaker's five-kingdom approach

In comparatively recent times, the 1960s, Robert Whittaker put forward what we refer to as the five-kingdom approach (Fig. 8). He assigned all known species to one of five kingdoms, based on evolutionary relationships determined by how they looked, how they worked and their pattern of development. Essentially, he added an extra kingdom, the Fungi, to Aristotle's scheme stripped of its supernatural elements. There were the bacteria, the plants (Plantae), the Protista (uni-celled organisms), the animals (Animalia) and the Fungi. We encountered all these kingdoms, or components of them, earlier in the chapter.

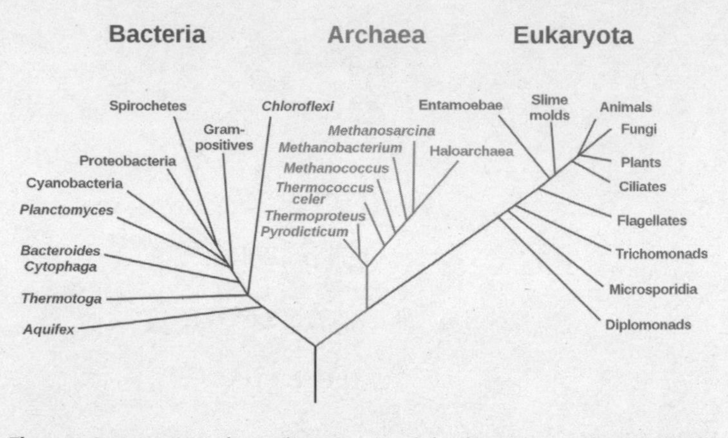

Figure 9 Woese's three-domain model. (Courtesy of Wikimedia Commons.)

Woese and the three-domain model

Even from its inception, the five-kingdom approach was the subject of much debate. To what extent did kingdoms actually reflect evolutionary relationships and to what extent were they artificial constructs? What was clear was that both Aristotle's and Whittaker's schemes emphasised *big* (principally animals and plants) over *small* (often microscopic living things) – understandable for Aristotle almost 2,000 years before the first microscope. But even despite the change in view wrought by Stanier and van Neil's two-domain model, the 'size-ism' inherent in the five-kingdom approach was not successfully challenged until the later 1970s, by the molecular biologist Carl Woese. Like so many micro- and molecular biologists he tended not to associate the importance or 'worth' of living material with its size. He devised a classificatory scheme based on the evolutionary relationships of common molecules (principally DNA and RNA, so a classification based on gene sequencing) to one another. What he came up with really shook biology,

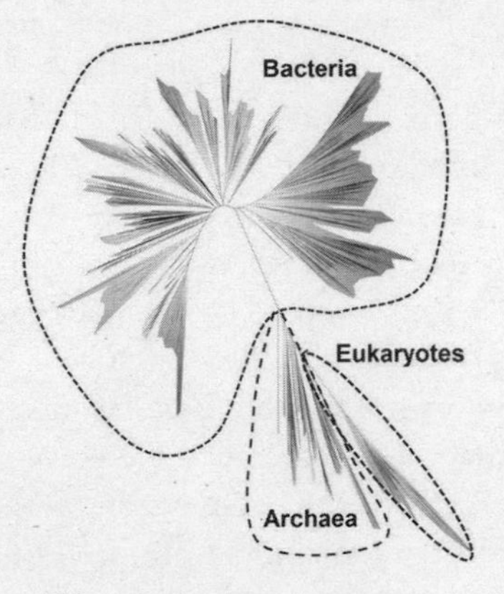

Figure 10 Hug's modified three-domain approach. (Adapted from Hug et al. (2016), *Nature Microbiology* 1, 16048.)

though it did not come as much of a surprise to the microbiologists. The three-domain model Woese proposed extended the two-domain model of Chatton by recognising that the Bacteria are in fact two very distinct groupings, termed the Eubacteria and the Archaebacteria (Fig. 9). Both are prokaryotic forms of life, but when Archaebacteria genes were sequenced about 50% of the genes were new. What is more, some of them resembled your genes and those of other eukaryotes more than they did those of the Eubacteria. So there were two very early branching incidents near the origin of life that resulted in the formation of three domains, the Archaebacteria, the Eubacteria and everything else (the eukaryotic organisms – the four remaining kingdoms if you like).

A new twist to the three-domain model

In the latest effort to explore what this 'tree of life' looks like, Laura Hug and her collaborators have used not just genes but entire genomes. Without going into detail, this latest tree shows a lot of evolutionary innovation in previously unknown or under-studied groups. What it also highlights that doesn't really appear in Woese's three-domain model is how all eukaryotes emerge from within the Archaean bacteria (Fig. 10)! It will be interesting to follow how this model develops as more and more genomes are produced, particularly those of currently under-represented groups. There still seems to be no lack of major surprises to come. For example, in 2018 Lax and colleagues investigated the molecular history of a soil organism (a hemimastigote) from Nova Scotia and concluded that although it has a eukaryote-like design, it doesn't really fit in any of the pre-existing categories, prompting them to propose a new suprakingdom home for it.

...and when is a tree a bush?

And finally, it's looking increasingly like the tree of life, lower down on the tree, is actually more like a bush of life (Fig. 11).

Figure 11 Horizontal gene transfer. (Source: Doolittle (1999), *Science* 284, 2124–2125.)

It seems there has been a lot of swapping of genetic material between the so-called simplest forms (a phenomenon termed horizontal gene transfer), but to say more is to go well beyond what is possible in such a short introduction.

Designs on life

The phylum and the Bauplan

The level below kingdom or domain is termed the phylum (plural phyla). Until recently we have thought of phyla as a set of distinct and easily recognisable body designs. The early German biologists who contributed so much to our studies of evolution and development referred to the design that characterised a phylum as the *Bauplan*, a German word that could be translated as 'blueprint' or 'body plan'. We could compare different phyla with different car makers, such as Ford, Peugeot, Nissan etc. Each company is in the business of making cars. Cars would be at the level of kingdom or domain. Each company has a suite of individual car designs (species) that fits under the broad umbrella of 'it's a Ford' (one phylum, characterised by the Ford *Bauplan*) or 'it's a Nissan' (another phylum, characterised by the Nissan *Bauplan*). After the species, the phylum has long been considered the next 'natural' and comparable grouping we have for classifying living things. On the whole, comparing a species of cat and a species of Fungi, or the phylum Arthropoda with the phylum Mollusca is arguably more meaningful than trying to compare two genera, two families or two orders.

However, cracks are beginning to emerge in this traditional view of how useful and meaningful the concept of a phylum is. It is proving increasingly difficult to form a watertight *Bauplan* for some phyla. Gene sequencing has thrown much new light on the evolutionary relationships between and within the current designated phyla. What a century ago was a largely sterile and irresolvable debate, about the origins of and relationships between the phyla, is today a puzzle that has been solved, although not every piece is yet in place. Certainly, some current phyla seem to have more than one common ancestor (i.e. they are polyphyletic) – a major issue when a key part of being a phylum is being monophyletic, that is, having the same ancestor. Furthermore, genetic analysis, particularly of microbes, is making it increasingly difficult to use the phylum concept even for animal groups. For example, recent molecular analysis has firmly lumped four seemingly very different marine 'worm' phyla in with the one pre-existing phylum, the Annelida ('true' worms). They were originally separated out because they each had a different *Bauplan*. As a result, the usefulness of this staple of biodiversity classification, the phylum, is now the subject of considerable debate.

Most phyla are not very species rich

We currently recognise 35–40 present-day animal phyla. Some, like the Arthropoda, are very species rich. But most are not. The same is sort of true for the 92 named major divisions (sometimes referred to as phyla) of bacteria, but here the classification is far from settled. Many bacteria seem to have a lot of similar features but often turn out not to be closely related.

As we've seen on the shore at Bird Rock, it is possible to find a large proportion of the present-day animal phyla. That is why this shore, and similar shores across the world, have long been used as zoological training grounds, introducing students to the diversity of animal life. But for some phyla or species you really have to look hard. It was at Bird Rock that I first saw a distinctive horseshoe-shaped ring of pinkish tentacles emerging from a sand-covered worm tube on the underside of a large rock low on the shore. It looked like the phoronid worm *Phoronis* sp., but as I was unwilling or unable (this is a protected area of coast after all) to remove it from the rock I cannot say for definite what species it was. It was definitely a phoronid belonging to the phylum Phoronida, a group with only fifteen recognised species worldwide, six of them occurring along the Californian coast. Why there should be so many different basic designs with so few successful, at least in terms of numbers of species, is still a mystery. In Chapter 4 we'll spend a bit more time discussing the origins of the major (mainly animal and plant) phyla and how they have fared through geological time.

An unequal distribution of life

In summary, no matter which level we look at, the kingdom (or domain) or the phylum or even below that, life is not equally distributed between the different groupings that have been categorised. Most biodiversity is contributed by a handful of phyla. Most phyla are simply not very diverse. Most animals are insects, and most insects are beetles. More than three quarters of all plant species are of the flowering variety. Most mammals are rodents. Taking all of animal life together, to the nearest approximation every animal on Earth is an insect. Notice my own animal-b(i)ased view? I can hear my microbiological colleagues whisper through a wry smile, 'Yes, but to the nearest approximation every living thing on Earth is a microbe.'

3

Where on Earth is biodiversity?

> Below the Mexican border, the water changes color; it takes on a deep ultramarine blue…the fishermen call it 'tuna water'. By Friday we were off Point Baja. This is the region of the seaturtle and the flying fish…the color of the Gulf littoral as a whole is distinctly tropical…there are but slight affiliations with animals of the north temperate zones…and comparatively few species are identical with those inhabiting the West Indian or IndoPacific areas.
>
> John Steinbeck & Ed Ricketts, *Sea of Cortez*

From Berkeley, south to the Sea of Cortez

Sol Light taught Zoology 112, Berkeley's core undergraduate course in invertebrate zoology (the study of animals without backbones) in the 1920s and '30s. By the late 1930s he had developed an extensive teaching programme. This was published in 1941 as *Light's Laboratory and Field Text in Invertebrate Zoology*. It was the key work for identifying marine invertebrates from the Central California coast, and has remained so down to the fourth edition published in 2007. It is an incredible achievement, indispensable to those who require a single volume for identifying marine life on the Californian coast. It does cover much of the marine life further south, on the coast around San Diego. However, Light's book has to be used with care in Southern California as a number of the more northern species are absent from San Diego, and in their place are southern species not mentioned in the text. When Ed Ricketts led John Steinbeck on an expedition into the Gulf of California in 1941, as recounted in their book, *Sea of Cortez*, he did so believing that even further south they would encounter a whole new world of marine life. Ricketts (together with Jack Calvin) had already produced his own, now iconic, guide to Californian sea life, *Between Pacific Tides*, and certainly had access to the syllabus that would form the basis of Light's book. But he also knew it would be of limited use where they were going. The identification guides he took with him were good for identifying some things, but not everything. While there were many new records, about 10% of the finds were undescribed species. Steinbeck and Ricketts also note that 'a considerable number of Panamic invertebrates…reach their extreme frontier in southern California.' The Gulf is now a UNESCO World Heritage Site, being recognised as one of the most diverse seas on Earth, with over 900 species of fish alone. Biodiversity is not evenly distributed along the Californian coast, which nicely illustrates the point that is central to the present chapter.

The types, and even number, of organisms present on Earth are distributed unevenly across the Earth's surface. The same species is not everywhere. However, there are distinct and recognisable patterns. We now briefly explore what we know of how biodiversity is distributed across the ocean and land masses; identifying hotspots, coldspots and gradients as we go from the poles to the equator, from the surface of the sea to its deepest depths, and from low-lying areas to the tops of the highest mountains. But first we start with a look at one of the strongest and most pervasive patterns there is by asking the question, 'How are the number of species in a given area related to the size of that area?'

And we start from where just about anyone could start – the rocky shore at Bird Rock.

More is more

Back to Bird Rock

Let's go to the flat rocks in the middle of the shore at Bird Rock that have just been uncovered by the outgoing tide – my last visit was with my son Ethan, who as a child accompanied me on so many shore adventures. In a very small area, say a quarter of a metre square, out on open rock surface, we easily find three small limpets, some tiny barnacles and a rough periwinkle. That didn't take long. We increase our search area to around a metre square. The limpets, barnacles and periwinkle are still there, but now they've been joined by at least two species of seaweed clutching on to the bare rock, and furtive sea anemones seemingly peppered with sand – and extremely wet if you accidentally sit on them – and a decent size chiton, almost indistinguishable from the rock it's nestling into – seven species. The beauty of these strange and unfamiliar forms captures our imaginations and fuels a growing compulsion to keep searching, to keep seeing. So we further extend our one metre-squared search to three. This larger area contains a number of small seawater-filled depressions, little aquaria sparkling bright and glistening in the sunshine (I suddenly remember why – this is Southern California, not the west coast of Scotland). There is also one deep gorge in the stone that seems to be lined with some sand and gravel. All seven species we've already found are encountered again. This time they are joined by some stalked gooseneck barnacles, four more types of seaweed, including one beautifully red species, *Corallina*, which strengthens and protects itself by incorporating what seems like concrete into its body. And the little tide pools, packed full and brimming over with life, are actually difficult places to identify specific living forms. Life seems piled on life in a picturesque disorder of often colourful, always spectacular, overlapping and unfamiliar forms that draw out wonder from you no matter what they are called. But an hour's intensive searching in three small pools, gently moving weed aside or looking carefully between large animals and in submerged cracks in the rock, has uncovered two different sea anemones, at least one orange sponge and a creamy-coloured one, two flower-like plant-animals related to sea anemones, a fairly large flatworm, two shrimp species, at least three species of crustacean amphipods, at least one species of darting copepod, a ball of brittlestars (I think the same species), an additional three species of snail, a mussel, two different types of small white-tubed worm, two different box-shaped crabs, and at least five additional species of seaweed. A quick look-see in the sediment at the bottom of the gully – the tide is coming in already – reveals two more amphipod species and at least three different types of worm – thirty-two different species in all. Not bad for a quick look-see. If we had been able (with a small army) to sample larger and larger areas, so that we started to take in the dry land, the lower shore and the nearshore (where within the Ecological Reserve there is a kelp bed, a submerged cobble patch, a sandy shelf and even a submarine canyon, all with their own distinctive species), that number would have increased substantially.

The species–area relationship

By and large, as the size of an area increases, so too does the number of species found in that area. This is referred to by the technical, but deceptively simple, term 'the species–area relationship'. And this holds not just for Bird Rock and its surroundings but for other beaches, and for inshore and deep waters, woodlands, grasslands, freshwater streams, whole countries, islands and even regions and continents – just about whatever environment, and whatever scale, you can think of. Most of our studies are based only on animals or plants. However, in 2004 Claire Horner-Devine and her colleagues demonstrated for the first time that bacteria also show a species–area relationship over a scale of centimetres to hundreds of metres in a New England salt marsh. And Thomas Bell and his co-workers found that the numbers of different bacteria were greater on big islands than small ones. Thus, bacteria seem to follow the same pattern as animals, plants, algae and fungi.

The species–area relationship is one of the strongest relationships we observe in biodiversity studies. To many, I suppose, this relationship seems intuitive – but if you think it through, it doesn't necessarily have to be that way. Why must animal, or plant, or microbial species be unevenly distributed in space? Interestingly there is still no consensus amongst scientists as to why we should get such a relationship. There are a number of different ideas but few of them are mutually exclusive.

There are two ways in which species can be added to any particular area. They may come in from other areas as immigrants, or else the species already present, for some reason, start to throw off new species, to speciate. And there are two corresponding ways in which species can 'disappear' from that area. They may leave (emigration) or they may become extinct. Put simply the total number of species in an area must depend *to some extent* on the relative importance of each of these four features and their interactions, and so it is here we begin our search for the explanation(s) underlying the species–area relationship.

Current opinion holds that when we are dealing with very large areas, much larger than Bird Rock and the surrounding area, on the whole immigration and emigration become less influential. If this is true then it is the balance between speciation and extinction that is critical in determining the relationship.

But it's equally likely that the species–area relationship may just be the result of larger areas containing more different types of environment, each with their own species that *fit* that environment. And what works on a local scale could also take place at a number of different scales – the La Jolla region with a multitude of different habitats and environments, Southern California, the whole West Coast, the entire US, and so on. Certainly, this was how the distribution of salt-marsh bacteria referred to earlier was explained.

It is not unreasonable to think that the explanations given could account for the species–area relationship. Then again there is also the unpleasant possibility that this 'relationship' is not actually a relationship at all. The fact is, the bigger area you choose, the more samples you have to take, and the more samples you take, the greater chance you have of finding new things.

So, to summarise, it seems that the larger the area, the greater number of species present, and this is because of: a) the balance between immigration and emigration, and speciation and extinction, with the last two probably being more important in very large areas; b) the greater number of different habitat types and environments in larger areas. All that said, over and against this pattern there are a number of fundamental differences between different areas. We will start with a truly global-scale difference, that between the ocean and the land.

Those who go down to the sea in ships

Two thirds of our planet's surface is covered with water, salty water. Furthermore, three quarters of the ocean floor is abyssal, that is, it lies at a depth of between five and eleven kilometres. This deep-sea environment is in total darkness, and is slightly colder than the inside of your refrigerator. As well as the large surface area of the sea bottom, the overlying water is potentially living space for life. So it is perhaps surprising that about 98% of all marine animals and algae are bottom living, with only 2% suspended in the water column, and even then they are restricted mainly to the upper sunlit layers.

In some ways, living things in the ocean are more diverse than those on land. Of all the body plans (phyla) that characterise current life on Earth, at least two thirds are restricted to the marine realm. There is only one animal phylum, the velvet worms (Onychophora), that is restricted to the land. And this greater diversity seems to hold even for lower classification categories. For instance, nine out of ten classes are marine. It's only when you come to the level of species that things are not so straightforward. Fewer than one in six described species are marine. This could mean one of two things. It could be that the numbers of species in the ocean actually do outnumber those on land and that we end up with a skew in described species simply because it is much easier to collect land organisms compared with sea organisms (as we've said, 75% of the ocean is at a depth of greater than 4 km). The case has been made for the deep ocean being more diverse than, say, an equivalent area of rainforest. But this is based almost entirely on scaling up some localised measures of biodiversity that seem to indicate that deep-sea nematodes are hyperdiverse. Not everyone accepts the assumptions and calculations involved and the debate continues. The alternative view, which is not so hotly contested, is that there really are more species living on land than in the seas, despite the fact that the ballpark species numbers we have at present will probably change.

Whichever way you look at it, we have a major land–sea contrast. While there are a greater number of body plans (or phyla) in marine environments, the land has a much greater number of species. Why should this be? Lord Bob May put forward a number of suggestions that might help us. Life began in the sea. Therefore, it has been present in marine environments for a considerably longer period of time than it has on land. So, given more time, the early 'innovations' that gave rise to the major groupings we saw in the previous chapter all took place here. Second, land habitats tend to be more elaborate and more different from one another than marine habitats. Such environmental diversity is thought to promote biological diversity. Linked to this second point, it is believed that plant-eaters in the marine environment tend to eat just about anything compared with land-based plant-eaters, which tend to specialise, even to the extent that one animal species will feed off one plant (or even just particular parts of one plant) species. Such specialisation may be linked to the throwing off of new species – where the descendants of one species are sufficiently different from the ancestor to be given their own name – but as a word of caution it is difficult to see which one is cause and which one is effect. Finally, land life may tend to be bigger than sea life on average. For example, photosynthesis on land is carried out by grasses and trees, while in the ocean it's carried out almost entirely by microscopic life. As it's supposed to be easier for smaller things to get around, the idea is that the small things should not throw off new species so readily. All of these are neat ideas but that is still not the same as actually knowing why there is such a profound difference between marine and land biodiversity.

Hotspots: A tale of two definitions

Arguably, one of the main reasons that the study of oceanography and marine biology is so popular in California is the richness of the marine life on the shore and in the coastal waters. There are also areas of very deep water close to the shore. This has been a constant draw for scientists and naturalists for over a hundred years, as the establishment and success of, amongst others, Scripps Institution of Oceanography testify to. It would not

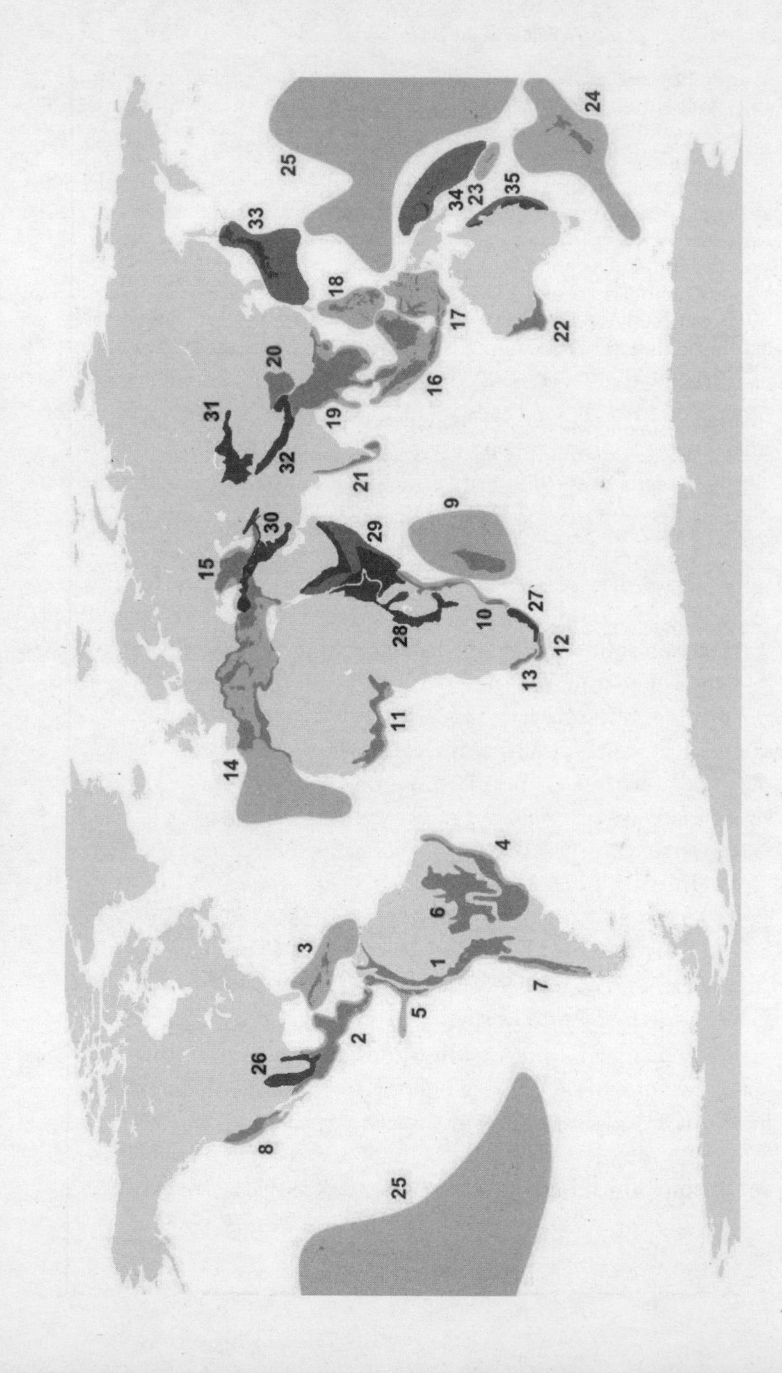
25
33
34
23
35
24
18
17
22
20
31
19
16
32
21
9
30
29
15
27
10
28
12
13
11
14
4
6
3
1
7
5
2
26
8
25

Figure 12 The twenty-five biodiversity hotspots designated by Myers and colleagues (2000, *Nature*, 403, 853–8), with ten additional sites that have been added since then (Williams et al. (2011), *Biodiversity Hotspots*, 295–310). The original twenty-five: 1) the Tropical Andes; 2) Mesoamerica; 3) the Caribbean Islands; 4) the Atlantic Forest; 5) Tumbes-Chocó-Magdalena; 6) the Cerrado; 7) Chilean Winter Rainfall-Valdivian Forests; 8) the California Floristic Province; 9) Madagascar and the Indian Ocean Islands; 10) the Coastal Forests of Eastern Africa; 11) the Guinean Forests of West Africa; 12) the Cape Floristic Region; 13) the Succulent Karoo; 14) the Mediterranean Basin; 15) the Caucasus; 16) Sundaland; 17) Wallacea; 18) the Philippines; 19) Indo-Burma; 20) the Mountains of South-West China; 21) Western Ghats and Sri Lanka; 22) South-west Australia; 23) New Caledonia; 24) New Zealand; 25) Polynesia and Micronesia. And the additional sites: 26) the Madrean Pine-Oak Woodlands; 27) Maputaland-Pondoland-Albany; 28) the Eastern Afromontane; 29) the Horn of Africa; 30) the Irano-Anatolian; 31) the Mountains of Central Asia; 32) Eastern Himalaya; 33) Japan; 34) East Melanesian Islands; 35) the Forests of East Australia. (Courtesy of Wikimedia Commons/Ninjatacoshell.)

be unreasonable, in common parlance, to say that La Jolla is a real hotspot for marine biodiversity.

The fact that biodiversity is unevenly distributed means that there must be highs and lows, hotspots and coldspots. Coldspots could also be recognised at different scales. Some wetland areas have relatively few species, as do some polar regions. The term 'biodiversity hotspot' was introduced by Norman Myers in 1988. Initially he designated ten tropical forests as hotspots of biodiversity and this rose to twenty-five in 2000. However, in the use of the word hotspot Myers meant more than just that they were sites of high species richness. To him they had to have large numbers of species that were found nowhere else (endemics), but he also defined them in terms of how threatened they were. So instead of hotspot being a term for high species richness, merely describing a natural pattern, the term was wedded to conservation and identifying conservation priorities from inception. So calling La Jolla a hotspot for marine biodiversity is not necessarily the same as calling one of Myers's forests a hotspot...but it might be.

Most talk of hotspots in the scientific literature follows Myers's usage. In fact, there is now a formal definition of hotspot in that it must be an area which contains at least 1,500 species of plant, and that area must have lost at least 70% of its original habitat. Hotspots tend also to be defined using larger organisms such as mammals, birds and plants. In 2019, inventories of hotspots recognised 35 different regions worldwide (Fig. 12). Exact figures vary, but broadly speaking half of all plant species taken together and less than half of all terrestrial vertebrates (72% of all mammals, 86% of all birds, 92% of all amphibians) occur in these hotspots. The Madagascar and Indian Ocean Islands hotspot was singled out as an area with a very high concentration of plant and vertebrate families found nowhere else on Earth.

Big-scale biodiversity: Biogeographical and political regions

Numerous attempts have been made to divide the surface of the globe into areas that differ naturally in terms of their biodiversity. During the middle part of the nineteenth century, naturalists collected plants from all over the world. Both specimens and notes were brought together and analysed in a number of famous herbariums. From such information, in 1866 August Grisebach produced a global distribution map for plants. Andreas Schimper went a little further when he produced a more detailed map in which he grouped the main plant types according to the latitudes – polar, temperate, tropical – in which they occurred. Studies of plant distribution went hand in hand with, and became intricately linked to, the study of climate, presumed to be one of the key drivers of plant distribution.

On land

I have on my shelves a beautifully illustrated book entitled *The Geography of Mammals* (1899) by father and son team Philip and William Sclater (William is first author). They introduce the book saying, 'Let us...dismiss from our minds...the ordinary notions of both physical and political geography, and consider how Earth's surface may be naturally divided into primary regions, taking the amount of similarity and dissimilarity of animal life as our sole guide.' They divide the globe into six main divisions. This scheme was originally put forward by the father, Philip, in an 1857 essay on the distribution of bird species read before the Linnaean Society. The scheme has remained largely unchanged until fairly recently (Fig. 13). Alfred Russel Wallace, who discovered natural selection at the same time as Darwin, came to the conclusion that 'admitting that these six regions are not precisely equal in rank, and that some of them are more isolated than the others, they are in geographical equality, compactness of area, and facility of definition beyond all comparison better than any others which

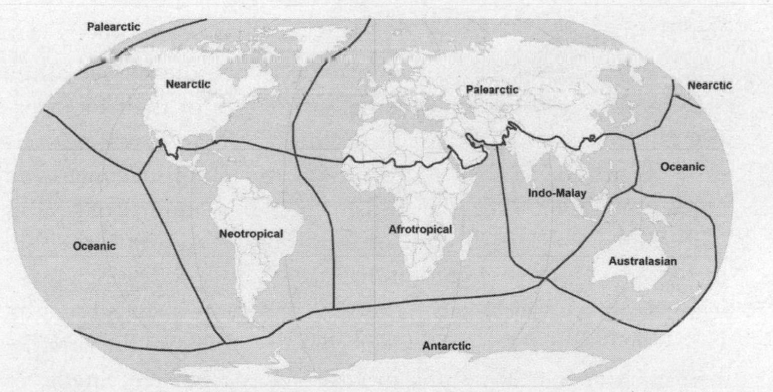

Figure 13 Generalised biogeographical regions of the world. (Source: Olsen et al. (2004), *Bioscience*, 51, 933–8, modified and extended at UNET-WCMC (2011).)

have been suggested.' In Sclater's scheme the land masses are split into six, what we now would term biogeographical regions. There are three tropical regions, the Ethiopian region (now the Afrotropics), the Oriental region (now the Indotropics) and the Neotropical region (the Neotropics). These three, however, are home to about 70% of land organisms. And of these three the Neotropics is by far the most biodiverse.

Relatively recently, in 2013, an 'update' of this scheme was proposed by Ben Holt and colleagues. Generated using information on the geographic distribution (and, for the first time, evolutionary relatedness) of over 21,000 species of amphibians, birds and mammals, they proposed an increase in the number of distinct regions to 20, grouped into 11 larger realms. This new scheme is still being vigorously discussed by the scientific community and it will be interesting to see how much traction it gains over the next few years.

Sea

Identifying natural divisions in the Earth's ocean has not been quite so easy. The reason for this is that there are fewer physical obstacles to organisms moving around in what is a much more three-dimensional environment than dry land. We know more about the surface of the moon than about the deep water which constitutes so much of our ocean. So it should come as little surprise that there is no agreed list of biogeographical regions for the world's ocean. That said, there are a number of large-scale patterns observable. The Indo-western Pacific area is believed to have the greatest concentration of marine biodiversity on Earth. It also has the greatest diversity of coral reefs and the organisms associated with them. Coral reefs are important for biodiversity generally as it has been estimated that these habitats are home to one in four of all marine species and one in five of all fish species.

In contrast with coral reefs, the deep sea (abyssal) has traditionally been considered poor in biodiversity. All of the major invertebrate and fish groups are present in the deep sea, but it is the echinoderms (brittlestars and sea cucumbers), not a species-rich phylum, that dominate. Also, while the mass of large marine animals is about 0.2 kg for every square metre of shallow sea floor, it falls to 0.2 g (a thousandfold difference) below a depth of 3 km. On the whole, more animals are found in deep water beneath the highly productive seas of polar regions than in deep waters in less productive temperate regions.

Biodiversity by country

Delineating biogeographical regions is our attempt to recognise natural barriers and discontinuities in biodiversity. However, most biodiversity information has not been collected and managed at such a mega-scale. Instead we have inventories collected at the level of individual countries. While it is not hard to see the reasons for this, information collected in this way must be treated with caution. Countries may be delineated on the basis of natural features. However, more often than not (as noted by the Sclaters), they are entirely human constructs – look at a map of Africa if you need convincing – with (depending on their size) little or no biological significance. Thus, while biogeographical regions may be the scale at which we want to understand and manage biodiversity, in reality it is often at the level of individual countries that scientific studies are initiated, decisions made, and conservation measures conceived and implemented. This being so, how is biodiversity distributed between different countries?

We think that most countries contain comparatively few species. About a dozen countries are estimated to contain one half to three quarters of the world's species. Most of these countries are in tropical regions and are amongst the poorest on Earth, an observation we'll return to. I said we think that most countries contain few species. That is because most countries do not have anything near a good, never mind complete, list of their resident species.

Latitude for life?

The land

Alexander von Humboldt (1769–1859) carried out some detailed investigations on the relationship between plants and their environment in tropical South America. One of his big findings was that changes in the types of plant found as you climbed up mountains were very similar to what you see if you journey from lowland tropical to polar latitudes. Since then others, including Alfred Russel Wallace, have observed that the number of species present in an area increases as you go from temperate (high) latitudes towards the equator (Fig. 14). This appears to be true not just for living organisms but fossil ones too. The same pattern even applies if we look at higher units of classification such

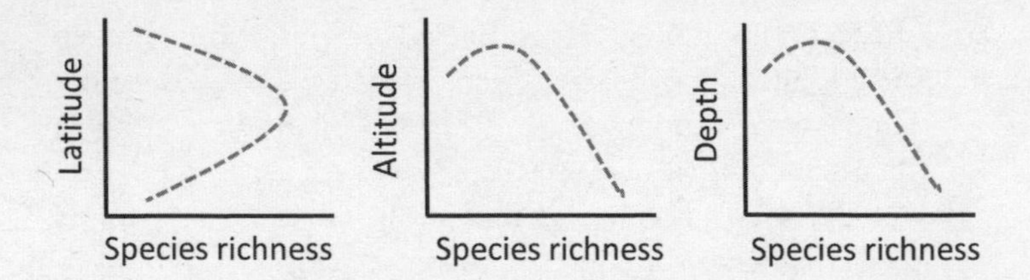

Figure 14 Three graphs showing the stylised relationship between species richness (as a measure of biodiversity) and latitude, altitude and depth.

as genera and families. Interestingly, this increase in species as you head towards the equator differs between the northern and southern hemispheres. Species richness increases more sharply in the former compared with the latter. So much so that some people have described the globe with its biodiversity as not so much like an orange but a pear shape. And even between species, some species are more 'tropical' than others, with one of the most well-known examples being butterflies which, for instance, are more 'tropical' than birds.

The sea

This pattern is very clear for land-based organisms, although slightly less so for marine. How much differences between latitudinal gradients of marine and land life are real and how much they have to do with problems in sampling marine life is not yet known. The pattern for deep-sea animals is clear – diversity increases as you head from temperate to tropical regions. Given our quantitative information for deep-sea burrowing animals, for example, comes from sample areas that in total are no more than one square kilometre, it is amazing we can detect any pattern. The pattern also holds for planktonic organisms, although it appears to be a little more complicated. It is only in studies investigating shallow-water environments that we find that the jury is still out. The evidence is conflicting. True, if we look at information on perhaps the best-studied group, coastal marine fish, they do show an increase in species richness from high to low latitudes, but it cannot be denied that the species richness of animals living on the seabed in Antarctic waters seems to be quite high. No matter what, the number of published studies that stop us talking about a general pattern for marine life cannot be discounted lightly.

Genetic diversity and latitude

In passing, some of the first work on how genetic diversity changes with latitude was published by Andreia Miraldo and her co-workers in 2015. Using publicly available databases for about 4,500 species of amphibians and mammals, they showed that genetic diversity increases as you head towards the tropics, being 27% higher in tropical compared with non-tropical regions. Conversely, genetic diversity decreases with decreasing latitude within a species of arctic-alpine plant, and no clear latitudinal patterns were detectable for marine viruses, or for marine and freshwater fish, although there was an association with temperature in the case of marine fish. Additionally, for the fish there was only a weak association (i.e. poor congruence) with species diversity. These recent genetic studies are timely and tantalising, serving to highlight how much we need more studies of how genetic diversity varies with latitude, and indeed, as we'll see below, with altitude and depth too.

Why is there a latitudinal gradient?

If the evidence for latitudinal gradients in diversity (at least in many, if not all environments) is relatively clear, the same cannot yet be said for why we see these patterns. Numerous explanations have been discussed. None of them are mutually exclusive and there are numerous versions and refinements of each in the scientific literature. First, it could simply be that there is a greater area of tropical habitats compared with temperate habitats just because of the shape of our planet. Also related to this 'problem' with the shape of Earth is the possibility that there may be more of the sun's energy reaching and available in tropical compared with temperate regions. More available energy, more species. Why it should follow that more energy means more species, rather than just more individuals of the existing species, is not clear, but there are even greater difficulties with this idea. Basically, when energy availability is actually measured for different latitudes and matched with species richness, sometimes there is a positive relationship and other times – well it's complicated and there can even be a hump-shaped relationship. Finally, it has been suggested that tropical animals have been exposed to more unbroken stretches of evolutionary time to throw off new species, fill habitats and so on. In this scenario the tropics were not so dramatically affected, and the evolutionary process was not interrupted, by such climatic or geological upheavals (e.g. drying out and glaciation) as higher latitudes. The only slight problem with this idea is that there is little consistent evidence that evolutionary rates are greater for tropical organisms. Furthermore, the tropical climate is not as stable over long periods of time as was once believed.

There is some evidence emerging that the present high species richness in the tropics is not a slow accumulation of new members over time, but in fact is due to a relatively recent 'outburst', in some cases as early as ten million years ago. No matter what, the latitudinal gradient in diversity must ultimately be produced (like the species–area relationship we met earlier in this chapter) by the balance between the rates of four different processes: speciation, extinction, immigration and emigration. At the large scales we are considering, immigration and emigration are likely to be small players. So in the scientific literature we find the tropics being portrayed either as a *cradle* (high speciation rates) or a *museum* (low extinction rates) *of diversity*. Currently there is good supporting evidence from fossil marine life for the *cradle* idea, but probably the truth is that latitudinal gradient in diversity is a balance between *cradle* and *museum*.

Altitude

Lessons from the tops of Scottish mountains

How does diversity change the higher up you go? When much younger, I climbed (scrambled up?) many of the Scottish Munros (mountains over 3,000 feet or 914 m) with a bunch of friends from Glasgow. One of the striking features of each climb was that you always set off surrounded by plants and grasses and trees, and an incredible host of insects, rabbits and birds. However, by the time you reached the summit you were invariably on hard rock, with the occasional lump of lichen or moss just clinging on, one or two tough-looking old birds hovering overhead (if you could see them through the torrential rain, mist and cloud) and no midges! It wasn't a quantitative survey, but I would have said that biodiversity definitely decreased as you ascended. And while I love the Scottish mountains passionately, they do not really compare in terms of height or scale to, say, the Andes, the Rockies or the Himalayas. And yet a mere 914-metre climb in Scotland can easily result in a reduction in temperature roughly equivalent to the temperature difference between where I live in Plymouth in the UK and Madrid in Spain (a distance equivalent to the length of California). So when we trawl through the quantitative published studies, do we see a reduction in biodiversity with increasing altitude?

Biodiversity takes the hump with altitude

As early as the 1800s, Humboldt, whom we met earlier, had noted that the number and type of plant species changed as he ascended the volcano Chimborazo in the Ecuadorian Andes. And taking together all the quantitative studies that have been carried out since, the simple answer is yes, there is a relationship. Whether we're investigating bats in Peru or treehoppers in Columbia, the number of species decreases with an increase in elevation. However, for flowering plants in the Kingdom of Nepal or ants in Colorado, to take two examples, the pattern is a little more complicated. They both show a sort of hump-shaped relationship (Fig. 14). That is, there is at first an increase in diversity with altitude. Then, as you continue to ascend, the pattern is reversed and you get a decrease in diversity with altitude, but the number of species at the bottom of the mountain is always much larger than the number at the top.

Now, given what we've already covered on species–area relationships and the mechanisms underpinning latitudinal gradients in diversity, it may have already occurred to you that different altitudes don't all possess equivalent surface areas. So maybe the relationship we see with altitude is affected, or even shaped, by the fact that habitat area almost invariably always decreases with increasing altitude. A study of tropical South American land birds found that species richness decreased with altitude. But what happens if you take altitude-related differences in habitat area into account? Using the same information, but this time standardised for area, the relationship was altered slightly. Instead of a decline in species richness with altitude, there was a hump-shaped relationship.

The humped relationship is also present for animal species such as ground spiders in Crete, insectivorous bats in Costa Rica, and plant species across altitudinal gradients in the Himalayas, south-west China, Mexico and western Norway. It has been suggested that the underlying reason for such hump-shaped relationships lies in an idea we've met before – energy availability. The theory goes that energy availability may be greatest at intermediate altitudes. Though the pattern is strong it is not invariant as species richness of small non-volant mammals in tropical north-eastern Australia increased with increasing altitude, a finding the authors put down to changes in vegetation influencing climatic conditions.

Mountains as islands?

Another theory put forward to explain the reduction in species richness with increasing altitude (although it doesn't really explain 'the hump') has to do with the isolation of mountaintops and places of similar height. Travel between mountaintops is difficult unless you can fly. Thus, emigration and immigration are low. Such isolation seems to result in new species being formed but, at the same time, a much greater possibility of populations becoming extinct. So, the argument goes, the balance between these two processes results in low species richness and the presence of many local (endemic) species at altitude. Why extinction rates should be greater with isolation is not all that clear. While the pattern is becoming clearer, as with the other big-scale patterns we still have a lot to learn about the underpinning mechanisms.

Aerial plankton and organisms in flight

Before we leave how biodiversity changes with altitude, we could ask the question, 'What about organisms living at altitude but without the benefit of a solid surface beneath them?' In 1953 a mountaineer saw a bar-headed goose, flying at an altitude of about 9 km, soar over the summit of Mount Everest. But the bar-headed goose is not like a swift or a frigate bird, which can spend most of its life in flight. And even the aerial plankton, spiders and insects that can be collected at such heights do not actually reside there. But what about the microbes? To quote the first microbiologist, Antoni van Leeuwenhoek, 'There may be living creatures in the air, which are so small as to escape our sight.' However, most of what is deduced about aerial microbes is based on the assumption that the atmosphere is a *conduit* for microbes getting around rather than a *dynamic habitat* where they actually live. That said, this assumption is increasingly being challenged. It has recently been estimated that, given how long they reside high in the atmosphere, some microbe types must spend fifty generations 'living' an airborne existence. So for potential patterns in aerobiology, it may still be too early to say, but literally watch this space.

Depth

The short-lived azoic theory

In perhaps the first book ever fully devoted to marine studies, *Histoire Physique de la Mer* (1725), Count Luigi Ferdinando Marsigli recorded his investigation of animals living in the 'deeps' of the Gulf of Lyons, France. Similar findings were made at depths of around 25 m in the Adriatic in 1750. Sir John Ross is credited with bringing up the first truly deep-sea animal, a basket star from a depth of just less than 2 km in the waters of the North-West Passage, in the early 1800s (although in truth the same species had been formally described by 1762 from material sent to England from the Caribbean!).

A key figure in the study of the sea in the first half of the nineteenth century was Edward Forbes. From his early twenties he spent a considerable amount of time dredging for animals in the seas around Europe. On board HM survey ship *Beacon*, he dredged to depths of up to 420 m in the Aegean Sea. Based mainly on the results of this voyage in 1841, he put forward a hypothesis that the 'zero of animal life was probably about 300 fathoms [about 550 m], below which extended a lifeless or "azoic" zone.' Despite the fact that Forbes clearly thought he was putting forward a hypothesis to be tested, within a short space of time it had become accepted fact that diversity decreased with depth until a critical, quite shallow depth, below which there was nothing. This belief was relatively short-lived (despite views to the contrary in some textbooks) as deeper and deeper sampling became possible, and many underwater surveys in connection with laying the first submarine cables found life at all depths. In a book entitled *The North Atlantic Sea-Bed* (1860), there is a drawing of a brittlestar brought up from a rope that had been at a depth of 2,300 m. The voyage around the globe by HMS *Challenger*, making measurements of the sea and sea floor and collecting animals from great depths, finally brought the existence of deep-sea life, often bizarre and fascinating, to public attention. There was the feeling that biodiversity decreased with depth but such a relationship was not based on quantitative information and analysis. There were, however, still creatures in the deepest depths.

Out of our depth

And there is an awful lot of deep water. Approximately 361 million square kilometres, just less than three quarters (71%), of Earth's surface is covered by ocean, which contains nearly all (97%) of Earth's water. And as we said earlier, a further three quarters of that ocean is abyssal, that is, between 4 and 11.5 km deep. The life that exists on the bottom (the benthos) inhabits a 'normal' environment where the temperature is less than 5°C and the pressure is 401–1,151 atmospheres. So we have a reasonable knowledge of shallow-water life forms, but know next to nothing of the organisms inhabiting the ocean proper.

Consequently, only relatively recently have we been able to start to put together information to scientifically assess any relationship between depth and biodiversity. Even then the data are scarce. For many groups, both of invertebrates and vertebrates, species richness decreases with increasing depth (Fig. 14). For marine 'woodlice' living in the northern seas, 65 species were found in relatively shallow water. As you go deeper there is a progressive decline until 4 km down there are less than 10 species present. The same sort of pattern holds for fish, which halve in the number of species present over a 2 km depth range. Numerous studies have found a decrease in species richness with depth. However, sometimes we also see the hump-shaped distribution we encountered when discussing altitude. For organisms free-floating in the water column, the peak in the hump is between 1 and 1.5 km deep, 1–2 km deep for big animals living on the seabed, and 2–3 km deep for large animals actually burrowing in the seabed. Interestingly, in this case the size of the particles that make up the sea bottom may be more influential in determining species richness than the depth those sediments occur at. If allowance is made for different sediment types, or other aspects of environmental differences, even on a small scale, it is not uncommon to lose any relationship that has been noted between diversity and depth. Thus, our current understanding is that the relationship between diversity and depth is perhaps just a little more complicated than the initial idea that numbers of species decrease with increasing depth.

A journey to the centre of the Earth

I love the part in Jules Verne's book when, during their journey to the centre of the Earth, Professor Lidenbrock shows his nephew and companion, Axel, an underground ocean full of prehistoric sea creatures, an ocean skirted by huge mushrooms and other sorts of plants. Lidenbrock had discovered ancient life flourishing at great depths within the Earth. I loved the idea when I first read it. I remember thinking at the time it was a bit far-fetched. But as is so often the case, science fact can be just as strange and wonderful as science fiction.

Trogloxenes (what a great word!) are species which, though found underground, must periodically return to the surface – cave bats and some insects are like this. Troglobites, on the other hand, are species that obligatorily live their whole lives underground.

They often show dramatic specialisation. If animals, they tend to be blind, pigmentless and are used to going hungry for long periods. If not animals but something else, they must have ways of capturing energy, as photosynthesis used by so many green things is not available, simply because, literally, they live where the sun never shines.

Microbes are abundant in the continental subsurface of our planet, to a depth of 5 km. This region accounts for a significant amount of Earth's prokaryotes. It is not always easy to tell, but there is evidence that abundance and richness decrease with depth, sometimes logarithmically. Microbes have been found in sediment and sedimentary rocks at the bottom of the ocean, 2–2.5 km below the sea floor.

Staying close to home

Not all species are found everywhere. Some are. They are mainly microbial species. Very few animal species are found everywhere. The beautiful moon jellyfish, *Aurelia aurita*, appears to have a worldwide distribution, although genetic analysis points to there being as many as nine different 'types'. In fact, most species appear to be restricted in where they are found. Such a species is called an endemic and the restriction in where it occurs is referred to as endemism. We've already encountered endemism when we discussed hot and cold spots of biodiversity, and found that the number of endemics was used to help identify hotspots. At one extreme there are organisms that only occur in one lake, or on one mountain. For example, Crompton's Orcutt grass is found only in one vernal pool (small pools that fill during the winter rains and dry during the spring to become filled with flowers) in the Jepson Prairie of Solano County, California. All five species of Orcutt grasses depend upon vernal pool habitats in California for their continued existence. They are found nowhere else. They contribute to the 30% of Californian plants that are found only in California. Compare this with 1% of endemic species in the United Kingdom, which has an area about three quarters that of California. Generally, some very small areas can have a disproportionately large number of endemics – oceanic islands, such as Hawaii and the Galapagos, for example. Isolation, whether it's vernal pools, islands or mountaintops, is firmly linked with endemism, although there is also a distinct pattern where the number of endemic species in a given area tends to increase as one heads from temperate to tropical latitudes.

Despite everything that's gone before, it would be wrong to give the impression that the range size of an endemic is always small. For example, turkeys, when living free and not frozen in the supermarket, are confined to the Neoarctic biogeographical region, a not insubstantial area. And four out of five Australian plants and animals are endemic – creatures such as the koala and the red kangaroo – because of the isolation of the continent from Asia for tens of millions of years.

Congruence: The holy grail of diversity?

It should now be apparent that putting together a complete atlas of biodiversity would be a colossal, not to say impossible, task. In fact, even trying to work with indices of biodiversity, such as species richness, and documenting patterns with latitude, altitude and depth is in itself a huge undertaking. Much of our information is based on plant or animal groups, and when we refer to animals there is, unsurprisingly, a strong bias towards the more conspicuous groups, the birds and mammals. We know very little of the invertebrates and much, much less about microscopic life in all its forms. And yet even with the disparate information we have, common themes are beginning to emerge. With all the appropriate caveats and provisos, species richness does decrease for a reasonable number of quite different plant and animal groups as you a) head away from the tropics, b) ascend to great heights up mountains, and c) descend to great depths in the ocean.

Given such patterns, it is not unreasonable to believe that what is happening with one group of organisms could be very similar to what is happening to another group of totally unrelated organisms. Just as the human pulse can be used to tell you so much about the normal workings of the human body, because of the interrelatedness of the different body systems, could it be possible that patterns for one (or a number of) group(s) could tell you about other groups, including some that we may never have time to examine? This is what is referred to as congruence. Congruence would make possible an operational atlas of biodiversity and we could even *include* groups of organisms for which, because of time and financial constraints, we have no original information. And here indeed is the holy grail of biodiversity. Think of the huge implications of successful congruence for those responsible for monitoring the environment and its associated biology. It could alter beyond recognition how we construct, introduce and enforce conservation measures, and how we *use* biodiversity resources.

How far have we got? Attempts have been made to search for congruence between groups for which we already have reasonably good information. The results of these studies are not always encouraging but it is early days yet. Back-of-the-envelope type analyses do seem to highlight degrees of congruence between different groups, but often when the detailed study is carried out the outcomes are much more mixed. Some groups yes, some groups no, some groups don't know. In 1997 Williams, Gaston and Humphries published a study in the *Proceedings of the Royal Society* which was one of the first to attempt to describe the worldwide distribution of numbers (at the level of family, not species) of seed plants and tetrapods (four-legged animals with backbones; mammals, reptiles and amphibians). These are groups for which we have some of the best worldwide distribution information. Interestingly, there was a fair degree of congruence. On a map of the world coloured according to the total numbers of families of these groups found at that location, it was possible to see previously identified hotspots of biodiversity in Colombia, Nicaragua and Malaysia. Most satisfying was the gradient of colour on either side of the equator, indicating that there was a latitudinal gradient in diversity exactly as we saw earlier in this chapter.

So, overall, we are beginning to put together some of the basic patterns of biodiversity. What is also clear, however, is that even producing these patterns still requires a lot more work. We are only beginning to uncover the reasons why such patterns exist, and have a long way to go before we can use well-known groups to predict things about less well-known groups.

There is an ancient map of the Northern Atlantic drawn by Sigurður Stefánsson in Iceland around 1590. The British Isles, Iceland and Norway are instantly recognisable. Gronlandia, which presumably is Greenland, and Helleland (Baffin Island) are pictured as promontories of a large continent in the west. A number of the other locations in the Arctic seas and on this great western continent are much more difficult to place. Compared with the most recent edition of *The Times Atlas*, Stefánsson's map looks a little comical, basic in the extreme. And yet that map and its predecessors were used in some of the greatest feats of exploration carried out by the Scandinavian countries. Great things were accomplished even with such a basic map. The *Atlas of Biodiversity* we have at the present time, I suggest, resembles Stefánsson's map in a number of respects. It is basic. It is provisional. But you have to start somewhere. And as was the case with Stefánsson's map, there is no reason why it cannot be used to plan and to act upon. Let's face it

– Stefánsson's map must have worked to some extent. And if we think our knowledge of present-day biodiversity is incomplete and provisional, what of the area we turn to next – the origins of that biodiversity from its roots in deep time right through to the recent past?

4

A world that was old when we came into it: Diversity, deep time and extinction

Fossils of [redwoods] have been found dating from the Cretaceous era while in the Eocene and Miocene they were spread over England and Europe and America. And then the glaciers moved down and wiped the Titans out beyond recovery. And only these few are left – a stunning memory of what the world was like once long ago. Can it be that we do not love to be reminded that we are very young and callow in a world that was old when we came into it?

John Steinbeck, *Travels with Charley*

One every twenty minutes?

A now retired colleague of mine, Paul Ramsay, spent a lot of time in the high Ecuadorian Andes. He was interested in describing and preserving the biodiversity of these beautiful highland areas. One morning over coffee he showed me a picture of a plant in flower. It was brand new to science and had just been described from specimens found only at this one location beside a high alpine lake, Lake Luspa at 3,900 m. The plant is *Loricaria cinerea*. His tone changed as he told me that the area had recently been cleared by burning, and all of the plants destroyed. There was a moment of quiet before we continued on to another subject. Just newly discovered, *Loricaria cinerea* was now extinct. Lost for ever.

Three species disappear every hour, according to EO Wilson, an entomologist and populariser of science who many see as the founder of modern-day interest in biodiversity. Species become extinct. Sometimes species disappear from a local area, but still can be found in other parts of the globe (local extinction). For others, like *Loricaria cinerea*, extinction is global. The mass of biological information carried by the genes, lost for ever. But extinction is not new. And if we look back into deep time, at the history of past life on Earth, we see that the appearance and disappearance of species played a major part in forming present-day biodiversity. Perhaps learning about the history of biodiversity, and the role of extinction in that history, will help us to see present-day extinctions in a clearer light. So in this chapter we briefly look at the history of life on Earth, from its beginnings to the present day, paying particular attention to the key events, including extinction patterns, that helped to shape that history. At the end of the chapter and into the two that follow, we look more closely at the extent and magnitude of the effects that humans have had on biodiversity over the past quarter of a million years, and how today's biodiversity crisis compares with 'natural' extinctions in deep time. So, starting at the beginning...

A life in the year of...

Before diving into the major patterns of ancient life on Earth, let's pause for a moment. The current estimated age of our planet is 4,600 Ma (mega annum, millions of years) or 4.6 Ga (giga annum, billions of years). That's a lot of years. As humans we find it difficult enough to think in terms of tens or hundreds of years, never mind thousands of millions. Having used a humble volleyball and the Southern Californian coast to get some perspective on huge distances in Chapter 2, can we do the same sort of thing for deep time? The writer(s) of the first book in the Judaeo-Christian scriptures, the book of Genesis, contracted the creation of everything to six days, with biodiversity being produced in the last three. Quite a neat way of teaching – taking your reader from the unknown, or inconceivable, to the known, by means of something analogous, more everyday, and much easier to grasp. At least that was the view of St Augustine of Hippo (354–430 CE). He believed that creation took place in an instant. For him the creation timescale in Genesis was a literary device, to make the story more comprehensible.

Calendars are devices we work from, and with, all of the time. Most of us, I warrant, are reasonably good at thinking on this timescale. So what would the history of biodiversity look like, in broad brushstrokes, if we contracted our 4,600 Ma (or 4.6 Ga) history to occupy the same time as our calendar year?

The Earth forms very early on 1 January. In terms of clear life signs (i.e. definite fossils), we have to wait until the end of March before the first prokaryote (remember them? Chapter 2, p. 54?) cells appear. We have an even longer wait, through spring to the end of the summer, for the first eukaryotic cells. They appear at the end of August. Most of the year is over before we get the first definite animals, midway through November. Up until this point there has been next to no life on land. The first invaders, plants and arthropods, get there at the very beginning of December. Something really nasty happens on the lead-up to Christmas and nine out of ten species become extinct. And yet within a day, life is up and rolling again. Sure, many forms have disappeared, but others have taken their place. It is now the age, or should I say the fortnight, of the ruling reptiles, including the dinosaurs. These creatures dominate the air, sea and skies until just after Christmas, when they disappear. In the few remaining hours of the year, on Hogmanay (31 December for non-Scots), humankind appears on the scene. As for recorded history, it occupies the same amount of time as it takes you to count one, two, three, four, five, six seconds. With a couple of ups and downs through the year, overall biodiversity has increased dramatically from the end of March to New Year's Eve. Most of the history of biodiversity is solely the history of microbes, at least until mid-November, when it becomes the history of lots of microbes and some plants and animals.

Precambrian – before life?

A schoolgirl changes our understanding of life before life – but no one believes her

In the summer of 1956, schoolgirl Tina Negus visited Charnwood Forest in Leicestershire, UK, with her parents, inspired by her passion for geology and the desire to see Precambrian rocks for herself. While her family picked berries, she noticed a strange fern-and-feather-like fossil impression on the rocks in a nearby quarry. She told her geography teacher who did not believe her. A little later, schoolboy (now geologist) Roger Mason went rock climbing with friends in Charnwood Forest, where he came across a similar impression to that which Tina had found. Having told his dad, the pair approached geologist Dr Trevor Ford at the soon-to-be Leicester University and told him about the find. In the words of Trevor, he 'quite frankly did not believe the boy'. After all, the rocks in Charnwood Forest belong to what geologists refer to as the Precambrian, and at that time there were known to be very few fossils in rocks older than the Cambrian (541 Ma) – Tina's geography teacher could have told them that. In fact, before the second half of the twentieth century, the Precambrian period was still thought of by its older name, the Azoic (without life) period. However, having gone to see the rocks for themselves, they found the fossil of what looked like a sea pen, a creature still found half buried in shallow marine waters today, and indeed it looked very much like a thickened feather. (The idea that it was a relative of the sea pen has since been challenged, and some would now put it in its own extinct phylum, the Vendobionta.) It was named *Charnia masoni* after the schoolboy who was, at the time, believed to be the first person to find it, and not *Charnia negusi*, as it might have been.

Since then the number of Precambrian fossils found has increased dramatically. As we've already seen in the *year in the life of biodiversity*, the oldest definite fossil is of a prokaryote, contained in rocks that are a staggering 3.5 Ga old. The oldest definite eukaryotic cell is found in rocks 1.5 Ga old. In fact, all of the fossils more than 700 Ma old are attributable to microbial life. Anything older than that tends to cause controversy. There have been claims of fossilised worms, or worm burrows, in rocks more than 1 Ga old, but this has not really been accepted by the scientific community as a whole.

The garden of Ediacara

To date, the earliest animals we know about come from rocks that are just short of 600 Ma old: Precambrian rocks from Australia, Canada, England, Namibia and Russia. The discovery of this variety of Precambrian life was made initially in Ediacara, Australia. The Ediacaran period (635–541 Ma) is a critical point in the history of biodiversity. It marks the transition from a purely microbial world to one which is *visible*, where large, complex, and often shelled animals are common. The strange, almost plant-like shapes of many of the animal impressions in these rocks led to this 'window into the past' being referred to as the *Garden of Ediacara*.

When first collected, the impressions were interpreted as varieties of the different body plans (phyla) that exist today (Fig. 15). There were organisms that looked like jellyfish. And Mason's

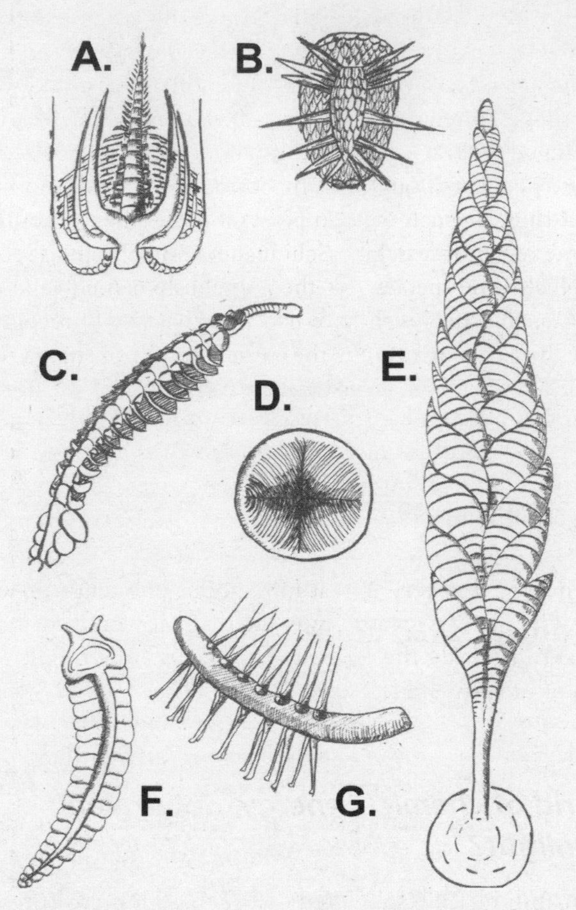

Figure 15 Examples of some Ediacaran (e) and Burgess Shale (bs) animals: A. *Marrella*[bs] (approximately 2 cm long); B. *Wiwaxia*[bs] (approximately 5 cm long); C. *Opabinia*[bs] (arthropod ancestor); D. *Dicksonia*[e, bs] (early flatworm or annelid worm – or is it even an animal? 18 cm long); E. *Charnia*[e] (3–60 cm long); F. *Spriggina*[e] (an animal, but not clear where it fits. It may be the ancestor of, or related to, the Burgess Shale's *Metaspriggina*, which appears to be a chordate similar to *Pikaia*, an ancestor of jawless fish. 3–5 cm long); G. *Hallucigenia*[bs] (1–5 cm long). (Redrawn by Ben Spicer from various sources.)

fossil too seemed to belong to the same group as the jellyfish and sea anemones (phylum Cnidaria). However, many of them were also quite *other*. All of the creatures were soft-bodied, and sported strange shapes, often flattened sheet-like, or even leaf-like, forms. They ranged from a centimetre to about a metre in size. What was more peculiar though was that initially there was no indication that these animals were in possession of either a mouth or a gut. Some recent research by Schiffbauer and colleagues seems to suggest that some species, like the gorgeously named tube-living worm *Cloudina*, may well have had what we would recognise as a gut. If they are correct, it is the earliest record of such a feature in the fossil record. Unfortunately, what we know of the biology of these creatures is speculative. The strange geometry has been interpreted in terms of the difficulties in getting oxygen into a largish animal that doesn't have either gills or lungs, or a circulation system to take oxygen to the tissues once it gets into the body. It is estimated that atmospheric oxygen at this time in Earth's history was very low, at most about one tenth of what it is today. The flatter a creature was, the easier it would be for oxygen to get into even the deepest tissues. Thus, one might expect animals to be thin and leaf-like.

A world of chemical energy, not driven by sunlight?

McMenamin in 1986 suggested that such gutless creatures must have secured their energy supply not by ingesting other organisms but by keeping within their tissues many tiny little algae. According to this view the algae were responsible for photosynthesis, capturing and transforming the sun's light into usable energy. Some of this energy went to keep the algae going and the rest was sequestered by the animal tissue. So they would be a bit like modern corals which have symbionts, photosynthetic algae, in their tissues. A major stumbling block for this idea is that photosynthesis requires light. Today, photosynthesis is restricted to algae either living in shallow waters or floating near the surface as part of the plankton. It is not a process normally associated with the seabed. Therefore, in 1989, Dolf Seilacher proposed that these hypothetical symbionts used not *photo*synthesis but *chemo*synthesis. In some parts of today's ocean there are animals that form an association with bacteria, and these bacteria use sulphur, extracted from water released from deep-sea hydrothermal vents, to produce energy. Why could the same thing not have driven this seabed ecosystem 600 Ma? A really nice story, but we still don't really know how Ediacaran animals worked.

How familiar is the Ediacaran fauna?

One of the key questions that emerges from careful examination of the Ediacaran fauna is, 'Could such strange forms really be attributable to animal designs (phyla, *Bauplans*) that exist today?' Some scientists, including the palaeontologist Martin Glaessner, have tried to interpret many of these rock impressions in terms of the phylum Cnidaria (which contains the sea anemones and jellyfish) and the phylum Annelida (the 'true' worms, like the ragworm, the earthworm or the leech). There was even a suggestion that Ediacaran creatures were not animals at all, but were instead large single-celled *microbes*. Others, such as Seilacher in the 1980s, thought that most Ediacaran organisms were animals but could not be assigned to modern-day phyla. They were, in fact, a very early 'failed experiment' in life – a bunch of *totally other* body designs that never made it through to the Cambrian period 541 Ma ago, never mind to the present day. If so, they were creatures that left no descendants. As such they tell us little about the rise of present-day biodiversity.

To be honest the jury is still out on what they are and how they worked. The final verdict may well have a bit of the 'unsuccessful experiment' view about it, but that is not a view shared by all palaeontologists.

The bottom line is that we know very little of bacterial, fungal or plant fossil biodiversity from its origins in deep time until comparatively recently. Undoubtedly, biodiversity increased from its inception until the end of the Precambrian 541 Ma. But the patterns are largely unknown. What we can suggest is that microbial life, bacteria and algae, dominated throughout this time, and so, because of the huge timescale involved, have dominated biodiversity for most of its history.

Explosive Cambrian

In sedimentary rocks that are about 541 Ma old, irrespective of where they are found in the world, we see the *sudden* appearance of many different kinds of fossils. They are the remains and impressions of all of the major animal phyla still present on Earth today, as well as a few phyla that do not seem to have left descendants. This is the beginning of the Cambrian (541–485.4 Ma, named after the Cambrian Mountains in Wales). The fossils are predominantly forms with hard parts, shells and armour, and the volume and variety is so impressive that this appearance in the fossil record is often referred to as the 'Cambrian explosion'. The very earliest Cambrian rocks, from places as far-flung as Siberia, China, India and Canada, contain a wealth of small – few greater than 1 cm in diameter – difficult to identify shells or skeletons, the so-called small shelly fossils or SSFs. Within a short space of time we find comparatively large animals appearing in the fossil record. In no time at all in geological terms we have life, and life in abundance. As yet all of this life is concentrated in the seas and ocean. There are no land animals or plants.

Cambrian forms

When we looked briefly, two chapters ago now, at some of the key characteristics of the most species-rich groups, we had to omit what might be regarded as major phyla. So, as we introduce each of the major animal phyla that make their appearance in Cambrian rocks, if we haven't encountered them before, we'll spend a little time providing some additional detail on the modern-day forms (Fig. 16).

1) Trilobites and crustaceans
As is still the case today, the joint-legged animals, the arthropods, dominated Cambrian seas. But the most common forms,

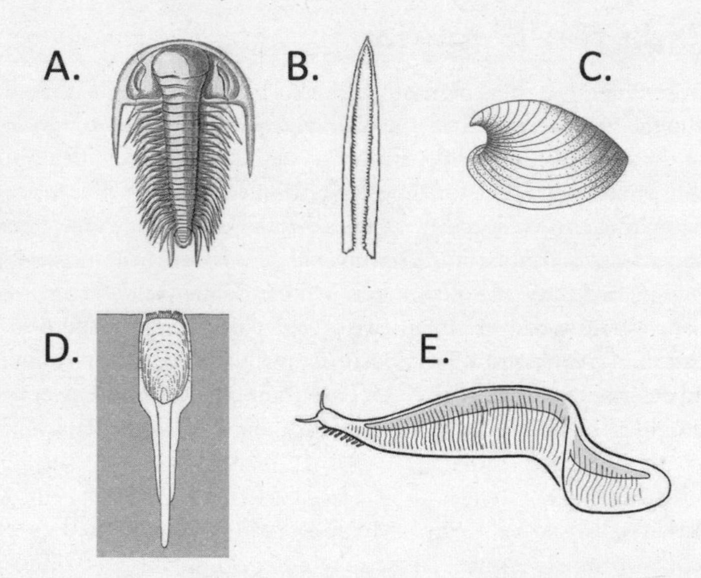

Figure 16 Common animals of the Cambrian explosion: A. Arthropod trilobite *Paradoxides* (10 cm long); B. Graptolite *Didymograptus* (6 cm long); C. Mollusc *Helcionella* (less than 1 cm long); D. Brachiopod *Lingula* (shell 2 cm long); E. Lancelet (Chordate) *Pikaia* (body 3 cm long). (Redrawn from various sources.)

the phylum Trilobita, are not creatures we would immediately recognise, mainly because trilobites have been extinct for a very long time – around 250 Ma. They look a bit like giant woodlice (Fig. 16A). In fact, trilobites are more closely related to scorpions and spiders than crabs or woodlice, and although they retained that basic woodlouse shape, they showed a tremendous variety of forms, often displaying different degrees of spininess. Crustaceans too make their first appearance in Cambrian rocks. While definitely sharing the crustacean *Bauplan*, they look quite different from the shrimps, crabs and lobsters of today.

2) Graptolites

Another dominant, but unfamiliar, group is the graptolites. More like bizarre etchings than animal impressions, these twig-like creatures appeared midway through the Cambrian (Fig. 16B). They were colonial animals living in connected tubes. Graptolites are now extinct (mostly by the end of the Devonian (419.2–358.9 Ma), although some held on into the Carboniferous (358.9 – 298.9 Ma), but they are related to what is now a very small group, with only a hundred living species, in the phylum Hemichordata. There are a couple of shelled species that still resemble graptolites, although modern-day forms are mostly wormlike (acorn worms), and look nothing like their ancestors. There are only ninety-six species of acorn worms worldwide, with twenty-three occurring on the west coast of North America. Two species from La Jolla have been found and described as recently as 2013. They must be relatively common, though, as the larvae can, at times, be very abundant in the plankton.

3) Molluscs

More familiar is the phylum Mollusca, snails and armoured cuttlefish. We've already encountered them in Chapter 2 as one of the most species-rich animal groups currently. Some of the earliest molluscs look a little like the periwinkles that we have today, with one of the most common species, *Helcionella*, more resembling a limpet (Fig. 16C). The armoured cuttlefish is basically a soft-bodied form living inside a small curved shell. Clams and their relatives do not really get going until after the Cambrian.

4) Brachiopods: Shelled but nothing like clams
Brachiopods look very much like clams, often on stalks (Fig. 16D). But they are quite different inside. Most of their body is made up from a multi-coiled, intricate and often beautiful feeding organ, the lophophore. Although they were big news in the Cambrian and later – there are 12,000 fossil species – today the phylum Brachiopoda is made up of a mere 300, mainly deep-water, species. Brachiopods are now but a shadow of their former selves in terms of their contribution to biodiversity. This said, they can still be extremely abundant. There are some beaches on the Namibian coast that run for miles and are made up almost entirely of the shells of dead brachiopods. *Glottidia albida*, one of only five species occurring on the Pacific Coast of North America, is found just below the low-water mark to a depth of 150 m, burrowed upright in soft sediments anywhere between Monterey Bay and northern Mexico. It is near identical in outward form and ecology to the lamp shell *Lingula*, which lived in the intertidal zone of Cambrian beaches 500 Ma.

5) Chordates: Biodiversity gets backbone
Until recently it was thought that vertebrates, backboned animals belonging to the phylum Chordata, did not originate until after the Cambrian. Certainly, there are numerous fossil lancelet species in Cambrian rocks from China and Canada, which shows that chordates appeared relatively early on the scene (Fig. 16E). But less than a decade ago the oldest fossil fish with a backbone was recorded from lower Cambrian rocks in Chengjiang in China.

Archaeocyatha: The only extinct phylum?

There is one phylum consisting of reef-building creatures, similar to sponges that were abundant at the beginning of the Cambrian but never seem to have made it to the end, the Archaeocyatha. They have the dubious distinction that they are the only phylum that, by common agreement, has become extinct. There were no sea urchins, sea lilies, brittlestars or starfish in early Cambrian seas, although there were creatures called Edrioasteroideans (now extinct) in abundance and they did a good sea urchin impression. There are also fossilised remains of some of the more soft-bodied phyla – more on those shortly.

Why diversify now?

One of the key questions when thinking about the rise of (at least animal) biodiversity in the Cambrian is, why now? What was so special about Cambrian times that saw the rise of all the major groups we see today? Before this time, our present-day continents were actually bunched together and made up one supercontinent. Just before and during the Cambrian, this land mass was breaking up, increasing the area of shallow seas around the continents, creating new habitats. It is in shallow seas that the Cambrian animals lived, as does much of present-day marine biodiversity. It is thought that the dramatic increase in shallow sea areas, ripe for colonisation, was accompanied by an increase in diversity. Some scientists have also suggested that the Cambrian was much warmer than today and/or that oxygen levels increased (due to the increase in the number and diversity of photosynthesising organisms) to a critical level, allowing the construction of relatively large animals. Yet others have drawn attention to the fact that much of the genetic machinery, *Hox* genes and the like, seem to have evolved around this time, and really the major genetic control systems of all animals originated and became relatively fixed in the Cambrian – there are no new (major) phyla after the Cambrian period. Ideas are plentiful but we still don't know for sure what kicked off this huge diversification event.

Cambrian explosion or short fuse?

It should be said that not everyone thinks that the Cambrian explosion was the big bang that it is traditionally made out to be. It has been claimed that there is a large, but for some reason unrecorded in the rocks, Precambrian history of many of the major groups. For instance, in 1997 Fortey, Briggs and Wills studied how Cambrian and recent arthropods were related to one another. They found that large tracts of time before the Cambrian were required to make sense of these relationships. For example, the ancestors of the millipedes and insects were suggested to have diverged from other arthropods before the crustaceans diverged from the chelicerates and trilobites. If this is true, as the crustacean divergence had occurred by the beginning of the Cambrian, this must mean that there is a long history for the millipede and insect ancestors that we know nothing about. More of a long fuse than a big bang?

Cambrian biodiversity: Good designs... or just lucky?

Given how key the Cambrian is to our understanding of present-day biodiversity (at least animal biodiversity and at the level of basic body plans, the phyla), it would be invaluable if we could visualise what the sea floor with all its animals actually looked like. Most fossil assemblages are higgledy-piggledy messes. But there are 'sets' of Cambrian fossils that show excellent preservation. They are believed by some to be crucial to the way we understand how biodiversity developed on Earth. So much so that their discovery and investigation formed the basis for a popular science book, *Wonderful Life,* written by palaeontologist and popular science writer Stephen Jay Gould. These fossils are collectively referred to as the Burgess Shale fauna.

How a small quarry in British Columbia changed our understanding of biodiversity

The Burgess Shale is made up of rocks 508 Ma old. It was discovered in the first part of the twentieth century by American geologist Charles Walcott, secretary of the Smithsonian Institution. He collected around 65,000 specimens during a decade of summer trips to a shale quarry (60 m long x 2.5 m deep) on the side of Mount Stephen in eastern British Columbia, Canada. Walcott interpreted the fossil animals he found against the backdrop of familiar body plans, of living phyla, much in the same way Glaessner had done when trying to classify the strange creatures he encountered in the Ediacaran fauna: if it looked like a trilobite then it was a trilobite and so an arthropod, and if it looked like a worm then it belonged to one of the known worm phyla. In the 1960s a palaeontologist from Cambridge, Harry Whittington, and some of his students re-examined, and to some extent rediscovered, the Burgess Shale animals (Fig. 15). What was so amazing was the excellent preservation of the specimens. Even soft-bodied forms, and lots of them, were preserved in exquisite detail. While many of the forms were familiar, there was much that was difficult to shoehorn into existing phyla

Remarkably, just as locally at Bird Rock, and globally for present-day biodiversity, the dominant group in the Burgess Shale was the arthropods. And of those arthropods the most common form was a trilobite-lookalike less than 3 cm long, *Marrella* (Fig. 15A). Lookalike because, like many of the other arthropods, *Marrella* did not exactly fit the trilobite *Bauplan*. This is a common theme of the Burgess Shale. The 120-odd species we find in the Burgess Shale allow us to look at the patterning of life while the pattern is still being woven. The largest animal found was the half-metre-long predator *Anomalocaris*, which would have moved through the water by the beating of giant flaps on its sides. Worms were present, as were sponges, soft-bodied forms that do not normally fossilise. And then there were some pretty peculiar forms. It is difficult to know what to make of the aptly named *Hallucigenia*, for example, a caterpillar-shaped creature with seven sets of limbs and seven spines along its back (Fig. 15G). In all likelihood it is probably a distant cousin of present-day velvet worms, so not as strange as it might first appear to be. These legged worms have a phylum all to themselves, the Onychophora, and with a grand total of about eighty living species they are now terrestrial and live in tropical regions. Ironically, however, the Onychophora is the only living phylum not found in the ocean today. And then there's *Wiwaxia*, looking so much like a half-walnut with a load of spines projecting out from it (Fig. 15B), which has been classified variously as mollusc, annelid worm, flatworm and unknown. What is interesting is that the Burgess Shale is typical of Cambrian times. It is not an exception. There are Burgess Shale-type faunas recorded from the US, Poland, Spain, Australia and China, with the most recent find in China in 2019 being older (518 Ma) than the original Canadian location.

'It's a Wonderful Life'

Gould described the Burgess Shale as the most important fossil deposit ever found. He wrote *Wonderful Life*, retelling beautifully the discovery of the Burgess Shale and its rediscovery by Whittington and his colleagues. Gould's main message in the book was that the Cambrian seas threw up many, many animal designs – far more than we see today. Very few of those designs made it through the Cambrian to form present-day biodiversity. Gould uses the story of the Jimmy Stewart film *It's a Wonderful Life* as a vehicle for his message. In the film, all-round good guy Stewart wishes he had never been born and gets to see a radically different world without his influence. And Gould's message? If we were to rerun the 'tape of life' then everything would be different. There would be no guarantee that the success stories of the Cambrian would be the same success stories in the Cambrian rerun. Gould's book is eminently readable, plausible, and as a result is convincing. However, one of Whittington's doctoral students who worked with him on the Burgess Shale, Simon Conway Morris, wrote a response to Gould. In *The Crucible of Creation*, Conway Morris argued that while there is much that is novel about the Burgess Shale animals, Gould had made too much of that novelty. He suggested that if you did rerun the 'tape of life' the result would largely be the same. The best designs would always win out. Arthropods were successful, and would be successful again on a rerun, not because they were lucky but because they had hit upon a good design. Conway Morris extended his ideas on the inevitability of the best designs winning out in a subsequent book, *Life's Solution: Inevitable Humans in a Lonely Universe*.

To conclude

We can conclude that at the beginning of the Cambrian we see a huge increase in the types of design that characterised early animal life. All of the major animal phyla alive today make their first appearance. By the end of the Cambrian there was a stabilisation of many, but not all, of these new groups. That's not to say that biodiversity overall stabilised, however, as we shall see next.

Post-Cambrian: Tinkering with successful designs?

In one sense all of the major innovations in body plan and design had taken place by the end of the Cambrian. The rest was tinkering…but fairly impressive tinkering as we shall see.

Palaeontologist Jack Sepkoski, who started collecting dinosaur bones and fossils when he was ten, spent much of his academic life putting together the ups and downs of biodiversity over the past 600 Ma. One of the graphs he produced has become almost iconic for those studying how biodiversity developed through time. It expresses changes in the number of marine animals (or at least the number of families of animals without backbones) through geological time (Fig. 17). The pattern produced has come in for some criticism, and still does, but it is probably the best attempt yet. This is against the backdrop of very poor preservation of groups in the fossil record.

In some ways, working out how the major patterns of biodiversity change with time from the fossil record could be likened to reconstructing the story of Shakespeare's play *Macbeth* with only access to Act II sc. 2, Act III sc. 4 and Act V sc. 8. It's

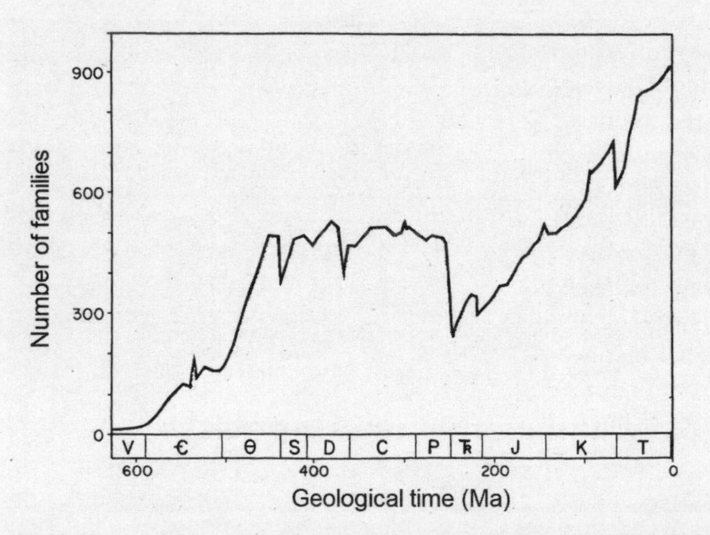

Figure 17 Sepkoski's graph of changes in marine biodiversity during the past 600 Ma. (Adapted from Sepkoski (1984), *Paleobiology* 10, 246–67, with permission.)

just about possible, but you certainly lose a lot of the detail. The problem, or major insight (depending on your point of view), that this analogy highlights is that if you were only left with Act II sc. 3, Act IV sc. 2 and Act V sc. 6, you would have the story of a drunken comedian with bad timing, performing his act just as a murder is discovered, followed by an unrelated woman (whose husband has just dumped her) watching her kid getting butchered – and it all ends in a big fight. The good news is that we also have such graphs for some marine protists, land plants, insects and four-legged vertebrates (amphibians, reptiles, mammals). Where they overlap with Sepkoski's graph they show broadly the same pattern. So we may not have much information, but what we do have, by and large, corroborates Sepkoski's big picture.

Palaeozoic – 'first life'

From Sepkoski's graph we can see that after the initial diversification in the Cambrian we lose a few families (small dip on the graph, even if more than that for those families) before there was an even greater diversification event in the Ordovician period (485.5–443.8 Ma). After the Ordovician ends, the number of families seemed to be stable – with bumps – for about a quarter of a billion years. That is not to say there was an absence of activity. The total number of people walking down a busy street may stay constant for some hours, but it is not the same people all of the time. In the same way, families present at the end of the Ordovician were a completely different set from those a quarter of a billion years later. It's just that new families were being added at the same time, and in the same proportions, as old ones disappeared. And this apparent stability also hid the fact that there were major innovations within the basic body plans that appeared back in the Cambrian. Cambrian fish were jawless forms. Now there were jawed forms, sporting a number of different fins, which we would recognise as the sharks and rays (jawed fish with cartilage skeleton and no swim bladder to keep them buoyant), as well as the bony fish. Fish diversity increased through the Silurian period (443.8–419.2 Ma), so much so that the following period, the Devonian (419.2–358.9 Ma), is often rightly referred to as the age of fish. The Ordovician also saw the first fungus in the fossil record, although they are thought to have first appeared a billion years ago.

Sometime in the Silurian period, or maybe just before, some of those bony fish, together with some arthropods and plants, made it onto land and established the first terrestrial ecosystems. Admittedly, early on this life was concentrated near the shore, but with advances in waterproofing both bodies and reproductive cells (spores/seeds), the land was not just colonised but conquered within a reasonably short period of time. Whether animals exploiting a new food source followed plants onto land, or animals came to prey on other animals ousted from the relatively crowded shallow seas, is still debated.

We get an insight into what early land life was like by travelling to the small town of Rhynie in Scotland, east of Aberdeen, and examining chert rock found there. The rock encapsulates, both in 3D and in exquisite microscopic detail, the plants and animals that formed these early terrestrial ecosystems. Silurian rocks contain an eight-metre-tall fossil, at first assumed to be a tree, but now shown to be a fungus. Without fungi, it is thought that plants may never have made it onto land. The earliest rootless land plants seem to have got access to essential water and minerals by an intimate association with filamentous fungi, one that transformed our planet. This critical association continues to the present day with fungi exchanging foraged minerals for the products of plant photosynthesis. The Age of Fish also saw the appearance of the first four-legged vertebrates, the amphibians, and the first insects. By the Carboniferous period (358.9–298.9 Ma) the land was very green indeed, with huge, very diverse forests. The remnants of this biodiversity has literally fossil-fuelled the world's economy for the past two hundred years. With the advent of some very tall 'trees', we see the appearance of flying insects, particularly dragonflies and some giant (1 m wingspan) dragonflies at that.

The Permian period (298.9–251.9 Ma) was a time of major climatic and geological upheaval. It is also the end of 'stability', and a fairly spectacular end at that. The half of all families lost at the end of the Permian that can be seen from Sepkoski's graph actually masks the true magnitude of this crash in biodiversity. There was a mass extinction of marine invertebrates with perhaps as many as 95% of all species disappearing. Some groups such as the trilobites and sea scorpions (aquatic relatives of spiders) disappeared for ever. The change in the character of animal life between the Permian and the period that follows, the Triassic (251.9–201.3 Ma), was so profound that it quite literally marked the end of an era, the Palaeozoic ('first-life') era.

Middle and modern life

The Mesozoic ('middle-life') era was the time of the ruling reptiles on the land, in the sea, and in the air. There was a new type of life in the ocean, with an expansion of shell-breaking predators and disruptive sediment movers. There was a marked increase in biodiversity, only slightly less rapid in its ascent than the diversification event in the Ordovician. Sometime in the Triassic period, or perhaps even before that in the Permian, the first mammals appeared – small, furry, nocturnal and highly secretive creatures. But it is the ruling reptiles that were in the ascendancy in the Triassic and dominated during the Jurassic (201.3–145 Ma). Originating from these reptiles was a flying dinosaur that gave rise to the birds. The Cretaceous period (145–66 Ma) witnessed the origin of the first flowering plants, the angiosperms.

The rise in biodiversity over the Mesozoic suffered a *small* setback at the end of the Cretaceous period. The small setback was, however, large enough to be seen as the end of this era, and the beginning of the Cenozoic ('modern-life') era (66–0 Ma). This is because many of the ruling reptile groups disappeared for ever, as did numerous invertebrates such as the ammonites. But, in terms of overall biodiversity, if anything the rate of increase accelerated greatly in the Palaeogene (66–23 Ma) at the beginning of the Cenozoic. At this point we are presented with a picture of biodiversity which has more in common with the present day than anything that has gone before. This was when there was a virtual explosion in the different types of mammals, birds, pollinating insects and flowering plants. In the Pleistocene (2.6–0.1 Ma), at the beginning of the Quaternary period, there were more different types of living things on Earth than at any other time in the history of biodiversity. At the end of the Pleistocene we come to the Holocene, where we begin to detect the effect of modern humans on the environment, and which, until recently, was seen as the epoch we now live in. But that could be changing. There's a new kid on the block.

The present – not set in stone

The new kid is the Anthropocene, a proposed geological era that could be termed *The Human Age*. It is characterised as a time of rapidly decreasing biodiversity, rapid climate shift, and the beginning of large-scale, highly organised activity by one dominating species – humans. The term was first coined by Paul Crutzen and Eugene Stoermer in 2000. They suggested that humans had influenced the planetary processes of Earth so much that the time we are living in should be formally recognised as a new geological era, the Anthropocene. It certainly caught the attention and imagination of artists, historians, philosophers, archaeologists, biologists, and also some groups who seem to just want to bash humankind for being so destructive, to name but a few…but not all geologists. Defining geological periods and eras is normally based on stratigraphy – the evidence that is found in layers or rocks, sediments and ice cores. The timescale of Earth's physical history is quite literally written in stone. So on which rocks is, or will, the Anthropocene be written, and what will they be marked with? The answer to that question lies in two other questions. First, is there enough *geological* evidence to warrant making a new *geological* era? And second, if there is, when did that era actually start? In terms of physical evidence, Colin Waters and colleagues point to markers such as the appearance of manufactured materials (plastics, aluminium, concrete); fallout from nuclear explosions, tests and accidents; particles associated with burning fossil fuels; and evidence of sea level rise in sediments. There is also a case for including climate change and perturbations in biodiversity (which we will consider in the next chapter), such as accelerated extinction rates and increased movement of species across the globe. Recently, the broiler chicken has been suggested as a living symbol of how biodiversity has been shaped by, and so is an indicator of, the Anthropocene. The look and internal workings of the bird is dramatically different from its ancestors of just fifty years ago. It is unable to survive without human help, and the combined weight of all the broilers exceeds that of all the other birds on Earth!

In terms of when, some would place the beginning of the Anthropocene as 1750, the date often attributed to the advent of the Industrial Revolution, while others claim that physical evidence of a shift in geology is only really obvious beyond the mid twentieth century, and that up until then any signals we detect are within the natural variability found in the *previous* era, the Holocene. The designation and timing issues are likely to run for some time, but I suspect that we will have a new geological era, and one that started demonstrably in the mid twentieth century.

The geological timescale we have just gone through, with some of the key events, is summarised in Table 1.

Table 1 Geological timescale and some of the key events in the history of biodiversity

Era	Period	Period begins (Ma)	Key events
[Precambrian*]		4,600	Origin of the five kingdoms/three domains of life, first many-celled organisms. Amount of oxygen in atmosphere increasing.
	Ediacaran	635	Soft-bodied organisms of the Ediacaran fauna.
Palaeozoic	Cambrian	541	Appearance in fossil record of all the major animal designs (phyla), including animals with hard parts – the Cambrian explosion. Jawless, heavily armoured fish. The 'Wonderful life' of the Burgess Shale.
	Ordovician	485.4	Rapid increase in marine biodiversity (no real land life, although the ancestors of land plants are definitely present). First fish with jaws appear – jaws and fins turn out to be a major innovation. Mass extinction (#1) at the end of this period.
	Silurian	443.8	First land plants and animals (arthropods).
	Devonian	419.2	First age of fish. First insects and amphibians (like salamanders, except some were about a metre in length) appear. Mass extinction (#2) at the end of this period.
	Carboniferous	358.9	Extensive forests of giant club mosses (40 m tall), tree ferns and horsetails (15 m tall) and first flying insects (like giant dragonflies the size of seagulls). Amount of oxygen in atmosphere 50% more than it is today. First reptiles appear (20 cm long) in Scotland, which is not yet associated with pubs.

Table 1 Geological timescale and some of the key events in the history of biodiversity (*continued*)

Era	Period	Period begins (Ma)	Key events
	Permian	298.9	A time of great upheaval – extensive volcanic activity. Origin of mammals – small furry, burrowing creatures. Mass extinction (#3) of marine animals towards the end of the period – nearly everything becomes extinct. Some animals, such as the trilobites, disappear for ever.
Mesozoic	Triassic	251.9	Ascent of the ruling reptiles (including dinosaurs). Fern-like plants dominant on land. Mass extinction (#4) at end of period.
	Jurassic	201.3	Ruling reptiles on land, in the air and in the sea. First birds appear.
	Cretaceous	145	First flowering plants appear, accompanied by a number of new insects (such as ants, bees and butterflies) which interacts with them (e.g. pollination). Mass extinction (#5) at end of this period. Ruling reptiles and many other groups, including the ammonites, become extinct.
Cenozoic	Palaeogene	66	The second age of fish.
	Neogene	23.03	The age of mammals, birds, snails, insects and flowering plants.
	Quaternary	2.6	Biodiversity peaks and begins to decline – many large mammals become extinct. First humans appear.
	Anthropocene		The age of humankind. The sixth mass extinction.

* Note that Precambrian is not an Era, it is a Supereon containing eight Eras.

Names of Eras/Periods and dates based on the International Chronostratigraphic Chart, updated by the International Commission on Stratigraphy (2020) (available at www.stratigraphy.org, or as an app for your phone – search for 'ICS timescale')

Most scientists believe that, even allowing for extinctions, biodiversity has increased from its origins to the present day. Exactly why that is the case is not clear, but it must be something to do with the balance between the origin of species and their demise – extinction (sound familiar?). These two components we discuss next.

Beginnings of evolution: The origin of species

New species can be thought of as accidents that happen as populations of individuals adapt to different, or changing, environments. Although Charles Darwin titled his book *The Origin of Species*, he did not actually explore the process of speciation very far, referring to it as 'the mystery of mysteries'. His emphasis was on showing that species change with time, not that evolutionary change necessarily results in new species. Since Darwin, considerable effort has been expended studying speciation, but ironically more attention has gone on studying the *products* of speciation rather than the *process* itself. This said, there are a number of things we can say about speciation in the context of biodiversity.

Many of the increases in biodiversity can be linked to times when new, empty living spaces appear. Someone has famously said that nature abhors a vacuum. As we've already seen in the Cambrian, the appearance of lots of new shallow-water environments was accompanied by an increase in diversity, in terms of new species and phyla. Interestingly, though there was room for diversity to flourish and many new species appeared in the wake of the Permian, there were no new phyla. The colonisation and conquest of land, too, was accompanied by the genesis of many new species and groups, but still no new phyla. Clearly, despite some recent claims to the contrary, there was something quite different about the Cambrian period.

In terms of the origin of species, it would appear that the easier the access to new and/or different environments, the greater the likelihood of new species arising – and that has happened throughout the history of life on Earth. The continual movement of land masses over the planet (plate tectonics), creating, multiplying, reducing or merging continents from the Cambrian to the present day, has on numerous occasions, and for a multitude of different groups, been associated with the appearance of new species. The isolation of Australia from Asia more than ten million years ago resulted in a whole suite of new species – 80% of Australia's wildlife is endemic. The development of slightly smaller scale barriers, such as mountains, hills, valleys, streams, lakes and channels, has also acted in the past as a successful contraceptive, preventing different populations of the same species from reproducing with one another. So two populations go their own evolutionary way and become quite different from the ancestral species.

Sometimes members of a population cross or in some way overcome an existing barrier, and they form a new population which may eventually lead to a new species forming. This is known as the founder effect. A good example of the effect is the 800-odd fruitfly species that occur on the Hawaiian Islands. Most are restricted to a single island, but genetic analysis tells us that the closest relative of a species on one island is often a species on a nearby island, and not, as you might expect, another species on the same island. It looks as if here there has been not just one but a minimum of 45 'founder events'. Climate, too, or should we say change in climate, has been implicated as a major driver of evolutionary change in past environments. When new species are produced because of separation by a physical barrier, this has the technical name allopatric (*allo*, 'different'; *patris*, 'country') speciation. However, it's not just outside changes that can result in new species.

There are numerous internal changes that could also produce novelty and ultimately lead to new species. So a new species arises where there is no physical barrier – this is referred to as sympatric (*sym*, 'with'; *patris*, 'country'). The fact that the mutations resulting in variation are the raw material of evolution, of natural selection, means that from time to time novel features may appear in populations that may give them an advantage either in the environment they are in, or in a different one nearby. The high diversity of cichlid fish in East African lakes seems to be the result of sympatric speciation. Slight specialisation in different feeding behaviours of different groups, even in the same location, resulted in those groups no longer mating with others of the same species.

In terms of the timescale of speciation, Darwin put forward a theory of gradualism, which is still largely accepted down to the present day. The accumulation of tiny changes over long periods of time leads to divergence and the origin of a new species, and over even longer periods of time, new genera and so on. But not, it would seem, new phyla – at least not since the Cambrian. Eldredge and Gould suggested that the appearance of major differences may in fact occur over a relatively short timescale. They envisaged long periods when very little happened, punctuated by times of extremely rapid evolution. They argued that such a model (punctuated equilibrium, it was called) made better sense of some parts of the fossil record. Punctuated equilibrium is still the object of some controversy, but it has to be admitted that there are problems with the gradualist approach when it comes to explaining large changes.

So we have numerous ideas of the sorts of factors that may promote or create the accidents that result in new species. We also have examples of each. Where we're still not entirely clear is which mechanisms have been most important and how common they have been throughout geological and even recent time.

End of evolution: Extinction

It is alleged that one eminent professor used to start his lectures with the phrase, 'To the nearest approximation every species is extinct.' Not quite true or the study of present-day biodiversity would be very brief indeed, but it makes the point. More than 99% of all species that have existed on Earth are now extinct. Extinctions were known about even before the advent of the theory of evolution. Count Buffon (1707–88), in one of his books, talks of 'strange fossil bones…have been found… Everything seems to suggest that they represent vanished forms, animals that once existed and today no longer exist.' When we looked at patterns of biodiversity through time, we noted numerous 'bumps'. All these drops in biodiversity, which act as bookends to each of the geological periods like the Cambrian, Ordovician and so on, are extinction events. Thus, such events define the categories we use to measure geological time. This said, there were five major mass extinctions.

The 'big five'

The end of the Ordovician may look like a little bump on Sepkoski's graph, but in fact this extinction event is estimated to have wiped out 85% of all species, profoundly affecting the trilobite, cephalopod, brachiopod and echinoderm groups. The end of the Devonian saw 75% of marine species, including nearly all of the trilobites and many of the coral reefs, disappear. For some reason, newly established life on land seems to have been unaffected. As we've already seen, the extinction event at the end of the Permian was the greatest so far. Many groups completely disappeared, including the last of the trilobites and the graptolites. Unlike the end of the Devonian extinction, both marine and land life were hit hard. Two thirds of insect families and just over two thirds of vertebrate families disappeared. The fourth extinction event was at the end of the Triassic period. Over a period of 15 Ma, three quarters of all marine species became extinct, and a good proportion of land species too. Finally, the fifth extinction event is one that we've already encountered, the Cretaceous–Tertiary boundary event. The main reason for its fame, even in popular culture, is that it saw the end of the dinosaurs and probably occurred as a result of a decent-sized meteor impact. However, both marine and terrestrial life generally were severely affected, with some groups, like the ammonites, disappearing for ever.

Causes of extinction

It would be good to know what brought about these extinctions. There are numerous suggestions: asteroid impacts, volcanic eruptions and other geological upheavals; deadly cosmic rays, alterations in ocean currents and growth of dead zones (no oxygen) in the ocean; and climate change. At present, it would seem most likely that the Permian extinction was brought about by climate change and the Cretaceous extinction by a decent-sized asteroid. The others? Pay your money and take your chance. Perhaps the most famous (and controversial) theory was proposed by Jack Sepkoski and David Raup, both at Chicago University. They proposed that mass extinctions were not random but took place roughly every 26 million years over the past 250 million years of life's history.

What may be of interest to us in later chapters is that the recovery of communities in the fossil record after these mass extinction events was in the order of 5–10 Ma. The players may have changed, but the number of players at this point is the same as it was immediately before the extinction event.

Extinctions as routine events in the history of life

These catastrophic changes in biodiversity could be compared with reshuffling the cards in the middle of a card game, eliciting major reorganisations in the way life looks and how things develop from that point on. This is probably true, although it is still not generally agreed. However, while much has been written about mass extinctions, in terms of overall loss of species they haven't really been that influential. The truth is that most species that have become extinct seem to do so by themselves. On average, species are only present in the fossil record for 5–10 Ma. Their rate of disappearance is referred to as the background extinction rate and it accounts for 96% of all extinctions. Both mass extinctions and background extinctions are almost routine events in the timeline of life. This said, the mass extinction event that could turn out to be really significant is the sixth one, the one that is taking place as you read this. And it's one that you and I are intimately involved in.

Early humans and biodiversity

There continues to be fierce debate as to the extent to which prehistoric societies impacted on biodiversity. By 10,000 years ago, and probably long before that, all of the major land masses showed signs of human occupation. There is also evidence that coinciding with the arrival of humans on a land mass was the disappearance of many of the large animals, particularly birds and mammals. This is perhaps nowhere more graphically illustrated than in the Pacific islands, where human invasion coincided with the extinction of many of the bird species, particularly if they were of the running (but obviously not fast enough) as opposed to flying type. Both hunting and modification of the original ecosystem seem to have taken their toll on the surrounding wildlife. Human influence may have been so dramatic, even early on in our own species's history, that we may never know what a natural ecological system actually looked like. When we talk of returning things to their natural state, the bottom line is that, in many cases, we have no idea what that natural state looked like. We can only surmise, and guess. This is no new revelation. Even more than one hundred years ago, Alfred Russel Wallace wrote, 'We live in a zoologically impoverished world, from which all the hugest, and fiercest, and strangest forms have recently disappeared... Yet it is surely a marvellous fact, and one that has hardly been sufficiently dwelt upon, this sudden dying out of so many mammalia [large mammals], not in one place only but over half the land surface of the globe.'

Extinctions post-1600s

Documentary evidence for extinctions improved from the 1600s onwards. So much so that we could begin to populate a graph of extinctions with time using contemporaneous data. But remember, almost without exception these were extinctions of generally large, easily noticeable animals and plants. Plants, birds and mammals figure large when we talk about extinction. The reason for this is simple. These are often the only groups we have good data for. The fact that it's difficult to get information for marine animals is borne out by the fact that only six marine species are known to be extinct. Similarly, we have a much better idea of what has gone on on islands than mainlands. Nearly three quarters of all mammal extinctions have been recorded from islands. Finally, all of this information is for described species – most are not. Whatever graph we put together is likely to be a huge underestimation. So, with all these caveats and qualifications, which

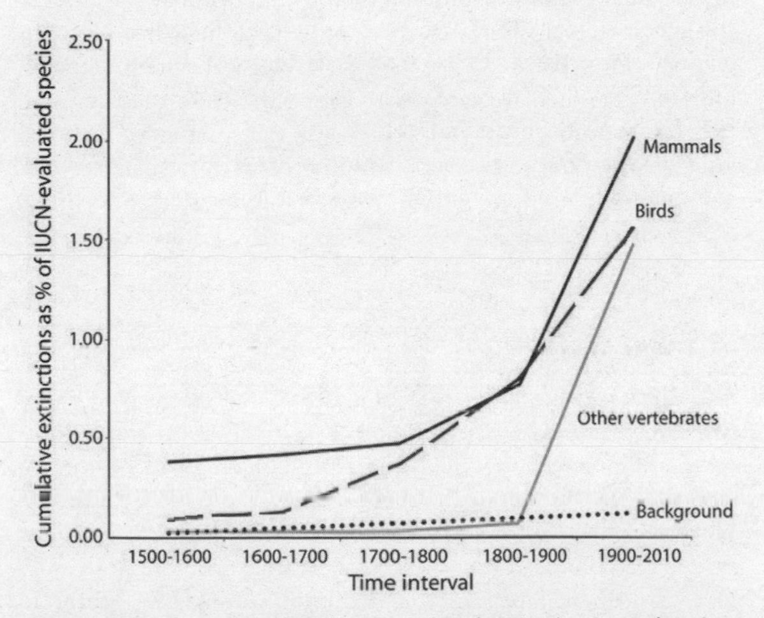

Figure 18 The rate of modern human-related extinction is accelerating. The graph shows how the number of extinctions amongst vertebrate species (mammals, birds and 'remaining vertebrates'), evaluated by the International Union of Conservation of Nature, has changed from 1600 to the present day. The dotted line shows the rate of background extinctions during this same period. Over the last century the extinction due to humans has been over one hundred times greater than the natural background extinction rate. (Adapted from Cebellos et al. (2015), *Science Advances* 1: e1400253.)

are so often the staple diet of science, Ceballos and colleagues have constructed such a graph for different animal groups. In Figure 18 you can see how the total number of extinct vertebrate species changes through time. It is clear that, even taking the most conservative approach possible, there have been more than 1,000 recorded extinctions since the 1600s, with more than half of those taking place in the last century. If we could zoom in on the graph, we might see a small dip after 1994 – but it isn't real. It is a consequence of scientists tightening up their definition of 'extinct'. From 1994 the International Union for Conservation of Nature (IUCN) defined extinct as 'when there is no reasonable doubt that the last individual has died', and extinct in the wild as 'when exhaustive surveys in known/expected habitat, at appropriate times, throughout its historical range have failed to record an individual'.

Proving extinction?

You don't have to major in biology to realise that it is difficult to prove that an organism is globally extinct. Local extinctions on the other hand are much more common and we have lots of examples we could draw on: the disappearance of skate from the North Sea, the loss of sturgeon from the Caspian Sea, a reduction in leatherback turtles in California from 212,000 to 40,000 in a thirteen-year period, the disappearance of 1,800 tropical forest populations every hour, and a decline in worldwide bird populations to one quarter of their numbers before the advent of agriculture, to name but a few. Perhaps our best-documented example is the whales, where many species are locally extinct and many endangered.

We said earlier that the average life expectancy of a species in the fossil record was 5–10 Ma. Taking information for birds and mammals the average life expectancy of a species today is, at best, less than ten thousand years. And we know that it takes 5–10 Ma for biodiversity to re-establish itself after a massive extinction event. All of the recent books and articles that proclaim the coming of a great biodiversity crisis are simply wrong. We are in the midst of a massive biodiversity crisis right now. What we always hoped was tomorrow has turned out to be today.

The Red Data Book

Each year the IUCN produces the *Red Data Book*, a compilation of all the information on organisms that are known to be threatened, extinct in the wild and extinct full stop. It is often referred to as the 'barometer of life'. Our knowledge of the global number of endangered species in key groups at this point in time (2020) is given in Table 2. In 2004 there were 15,503 endangered species. In 2020 the figure is almost double that. In the groups for which we have the best information (i.e. groups with the greatest percentage of described species evaluated – the vertebrates, the gymnosperms and the flowering plants) the figures are stark. A third of all amphibians, one in five mammals (although the *Red Data Book* notes it is more likely to be nearer one in four), and almost one in eight birds and reptiles are endangered. When it comes to the plants, it's over one third of the gymnosperms and almost one half of the flowering plants that are endangered.

Notice that while the information for the birds, mammals, amphibians and flowering plants on the list is pretty good, it is practically non-existent for most of the other groups, even common and species-rich groups that we looked at two chapters ago.

Table 2 Number of threatened species in 2020 compared with 2004

Group	Number of threatened species 2004	Number of threatened species 2020	Threatened as % of described species 2020	2020 % described species evaluated	% increase 2004–2020
Vertebrates					
Mammals	1,101	1,299	20.0	91	18
Birds	1,213	1,486	13.3	100	23
Reptiles	304	1,406	12.6	70	362
Amphibians	1,770	2,276	27.8	84	29
Fishes	800	2,849	8.0	59	256
Subtotal	5,188	9,316	12.9	73	80
Invertebrates					
Insects	559	1,819	0.2	0.9	225
Molluscs	974	2,275	2.4	9.0	134
Crustaceans	429	734	0.9	4.0	71
Others	30	591	0.2	0.7	1970
Subtotal	1,992	5,419	0.4	2.0	172
Mosses	80	164	0.7	1.3	105
Ferns	140	265	2.2	6	89
Gymnosperms	305	402	36.1	91	32
Flowering plants	7,796	16,667	45.1	11	114
Subtotal	8,321	17,498	4.3	10	110
Fungi	0	166	0.1	0.3	–
Lichens	2	27	0.2	0.2	1250
Protists (just brown seaweeds)	0	6	0.1	0.3	–
Total	15,503	32,432	0.2	0.3	109

Selected information from the 2020–2 *IUCN Red List of Threatened Species*. Threatened is defined as species that are Critically Endangered (CR), Endangered (EN) or Vulnerable (VU). N/A = not applicable.

Other takes on extinction

The biennial *Living Planet Report* takes a slightly different tack when presenting information on extinction. It introduces a number of 'Living Planet Indices' as a way of assessing our impact on the health of our planet. They do this by examining how 16,700 populations of mammals, birds, fish, reptiles and amphibians have changed over time. From these data they produce indices for specific areas or species, but they can also give a picture of what is happening globally (Fig. 19). Between 1970 and 2014 there was a 60% decrease in population sizes globally. These reductions were most marked in the tropics with an 89% reduction calculated for Central and South America.

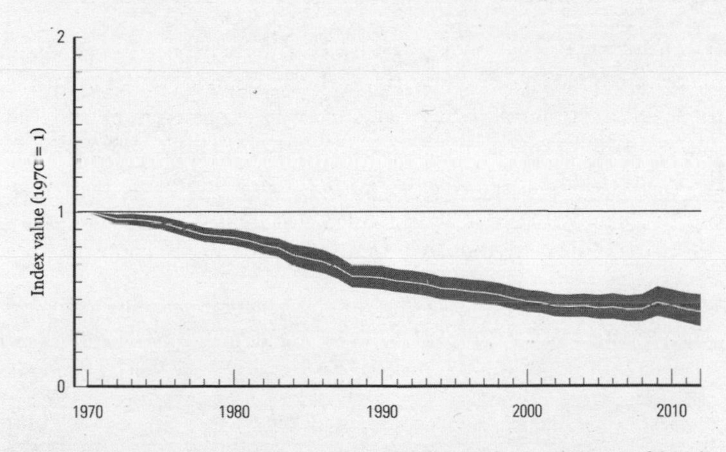

Figure 19 Global Living Planet Index (1970–2012). A measure of biodiversity, based on the changes in populations through time, of 14,152 populations of 3,706, compared with those same species in 1970. So, for example, an index of 0.5 would indicate a 50% reduction in those species since 1970. The shaded areas either side of the index line are confidence intervals (i.e. although there may be some variability around the index, the shading shows the region where we are 95% confident that we have taken that variation into consideration). (Source: 'WWF Living Plant Report: Risk and resilience in a new era', WWF International, Giand, Switzerland: CCBY.)

Attempting to take all of the uncertainty in these two approaches into account, the 2019 *Global Assessment Report on Biodiversity and Ecosystem Services* is the most comprehensive assessment of biodiversity to date. Published by the Intergovernmental Science-Policy Platform on Biodiversity and Ecosystem Services (IPBES), it estimated that 1 million species were potentially threatened with extinction, 'many within decades, more than ever before in human history'. This message was reinforced by the United Nations' *Global Biodiversity Outlook*, published just over a year later in September 2020.

To conclude

In summary, extinction is a routine and integral part of what makes up today's biodiversity. It has always been around. However, the level of extinction which we currently experience, extinction associated with our activities locally and globally, really is something else. So much so, as we've seen, the case has been made that we are now living in a new geological age, the Anthropocene. And it is to our activities that we turn in the next chapter.

5 Swept away and changed

> Animals may dig holes to live in; may weave nests or take possession of hollow trees... They make little impression on the world. But the world is furrowed and cut, torn and blasted by man. Its flora has been swept away and changed...its flat lands littered by the debris of his living.
>
> John Steinbeck, *The Log from the Sea of Cortez*

Threatening behaviour

San Diego has been touted as the most biologically rich county in the continental US...and the most threatened. Two hundred species of animals and plants are currently considered endangered, more than in any other county. If I asked you what the present threats to San Diego's biodiversity are, or even those to biodiversity in the area where you live, what would you say? Is it increasing development, change of land use leading to loss of habitat, increasing pollution, introduced pests, or competition for resources? And is this the same if we are thinking on a global scale? In this chapter we explore a number of direct (or proximate) causes (Fig. 20), going through them one by one, asking ourselves how significant each is. It will also be helpful to discuss the extent to which they explain past extinctions, and if

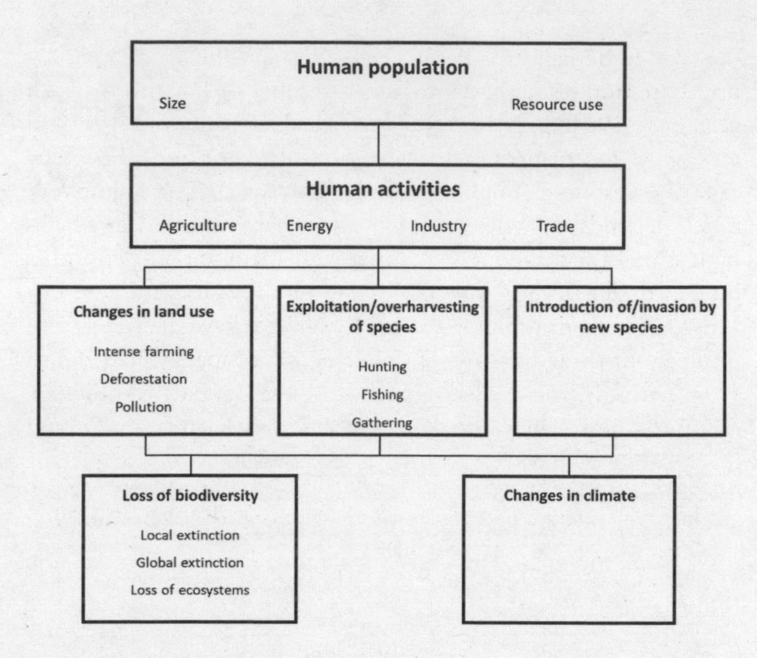

Figure 20 How we threaten biodiversity.

unabated, contribute to future ones. Whatever turn out to be the most important direct threats, we must look at what drives them in the first place, the indirect or ultimate threat. So we will finish with some discussion of the magnitude of the ultimate cause of current extinctions – that's you and me.

Living beyond our means

The 2016 *Living Planet Report*, published by the World Wildlife Fund, assessed our impact on the planet by examining what biodiversity is present and how we are using it. Figure 21 is based on that report. It shows our ecological footprint, the ratio between the world's demand for natural resources and the world's biocapacity (resource supply, expressed as number of Earths), for each year from 1961 to 2016. (You can calculate your own personal ecological footprint at www.footprintcalculator.org.) The biocapacity (resource supply) of Earth is always 1. The value very much depends on where you live and what you do. It can be as high as 3–4 for someone living in a high-income country, and as little as under 0.5 for someone living in a low-income country. You can see from the graph how globally we have changed from using about three quarters of the planet's biocapacity in 1961 to 1.6 in 2016. If true, we are currently at least 0.6 Earths in deficit. Assuming business as usual, we will be a full Earth over around the beginning of the 2030s. Given that there is only one Earth, this is not good news.

The Global Footprint Network, an independent US-based think tank seeking solutions to how to best manage our natural

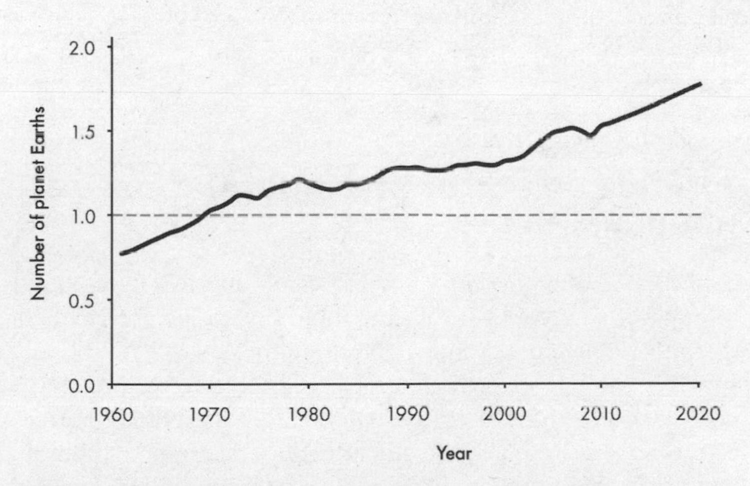

Figure 21 Global demand for natural resources, 1960–2020, comparing supply (dashed line) and demand (solid line).

resources and respond to global climate change, has come up with another way of highlighting how much our activities impact the planet. They proposed the idea of designating an 'Earth Overshoot Day' each year. This day marks the point each year when we go into deficit – the point at which we are using Earth's resources faster than nature can replenish them. And so 'Earth Overshoot Day' has been 4–6 days earlier each year from 1970 until 2019. In 2020 that changed – 22 August was the date, more than 3 weeks *later*. This dramatic shift was due to a 9.3% reduction in our ecological footprint, between 1 January and 22 August, compared with the same period in 2019. This reversal is a direct consequence of the COVID-19-related lockdowns globally. Unless something changes dramatically, there continue to be further lockdowns or there is more deliberate action on our part, it is likely to be a temporary blip, an ineffective speed bump.

No matter how we look at it, we are living well beyond our means in what was referred to in the 2005 *Millennium Ecosystem Assessment* as 'an unprecedented period of spending Earth's natural bounty'. That assessment reported that about 60% of the ecosystem services which support life are being degraded or used unsustainably.

Top five direct (or proximate) causes of biodiversity loss

The 2019 IPBES *Global Assessment* lists the top five most common known causes of animal extinctions since 1600. Remember that the extinctions we know about are just a small and biased subset of what is a largely unknown quantity, the total number of extinctions. And exactly how this relates to the total number of species threatened is also unclear. But these top five direct (or proximate) causes are still a good starting point for discussing extinctions and threats. The two most common causes of known extinctions are believed to be 1) habitat loss/restructuring and 2) direct exploitation. Habitat loss takes first place on land; direct exploitation first place in the ocean. Third is climate change, fourth pollution and then fifth the introduction of foreign, disruptive species to new environments. This priority will be used to structure the text that follows, except that pollution will be included with habitat loss and restructuring, under the larger umbrella title of habitat loss and degradation.

1) Habitat loss and degradation

Many species, and particularly tropical species, have specific food and habitat requirements. The more specific these requirements (their niche or, as the ecologist Elton put it, their 'profession'), and the more restricted their habitat (where they live, their 'address'), the greater the threat of extinction in the face of habitat loss and degradation. In the worst-case scenario, there just may not be anywhere else for life to live. So the only species to do well out of habitat loss would be those that are in some way pre-fitted for the new degraded habitat, or species that are generalists and really don't mind where they are or what they eat.

Nowhere on Earth is untouched by human actions. Restructuring nature is a common and persistent feature of our history and activities – agriculture, logging, industry, mining, pasture, roads, settlements and the like. Over the last thousand years we have moved more earth than has been mobilised by all of the natural processes put together. Fifty years of soil loss from our planet's surface would fill the Grand Canyon. Often what are commonly regarded as 'natural' landscapes are not. And the cost of such restructuring can be substantial. The global cost of habitat destruction was estimated about a decade ago at US$250 billion every year, with half of an ecosystem's economic value lost when a *natural* landscape is converted to human use. The 2005 *Millennium*

Ecosystem Assessment warned that about two thirds of the 'natural machinery' that enables life on Earth is being degraded, or used unsustainably, as a result of human action. When we talk of 'natural machinery', we mean things like fresh water, fisheries, air and water purification, the regulation of regional and local climate, natural hazards, and pests (we'll cover these ecosystem services in more detail in the next chapter). The 2019 *Biodiversity Assessment* estimates that it is now three quarters of global land surface that is significantly altered, and two thirds of the ocean.

The conversion of natural landscapes, particularly forests, for agricultural purposes has a long history. Beginning in China almost four millennia ago, it was near completed in Europe even before the time of the Industrial Revolution. Lowland forests began to disappear in the tropics of the New World at the same time as Portuguese and Spanish colonialism was in full swing (after 1500). Over the last two hundred years, considerable areas of land have been turned over to agriculture in the US and it was in the last century that we saw rainforests, in both North and South America, targeted. More land has been converted for agriculture in the last sixty years than in the eighteenth and nineteenth centuries combined. At present, about two fifths of the planet's land surface, or half of the habitable land, is cultivated and just over one tenth given over to cropland. Remarkably, in some parts of the US, land that had reverted from farmland to second-growth forests during the Great Depression saw the population of white-tailed deer explode despite intense hunting pressure.

We have destroyed or degraded between one third and one half of the world's forests. Interestingly, the 2020 *State of the World's Forests* report noted that there has been a bit of a slowdown in the rate of deforestation. Between 2015 and 2020, there was a 38% reduction in rate compared with the 1990s. There is good evidence that in some countries the area of forest (tropical to temperate) is actually increasing. Still, global forest cover worldwide has decreased by more than 80 million hectares since 1990. That's equivalent to just over 8% of the total US land area. Tropical forests in particular are so important because they are home to about half of all land biodiversity. They cover about 13% of the Earth's surface. Each year 140,000 square kilometres of rainforests are destroyed, for timber or farmland. While large areas of rainforest remain in South America and Central Africa, and, as previously mentioned, in some countries the areas are actually increasing a little, overall they are still disappearing, and areas where there are massive increases in human population, such as West Africa, contain some of the most endangered rainforest in the world.

And it's not just forests and woodland that are in decline. About half (more for tropical regions) of the grasslands or savannahs, one quarter of scrublands, more than four fifths of wetlands, and even about a tenth of the desert regions (hot and cold) show some sort of human disturbance. Certainly, some of the practices of intensive commercial agriculture – the destruction of hedgerows, the elimination of weeds and insect pests – have had a massive effect on biodiversity, particularly in higher-income countries.

While habitat destruction and degradation have been investigated in some detail, habitat fragmentation as a threat has not received the same amount of attention. An intact forest, woodland or meadow can be sectioned into small isolated bits by driving a road through the middle or effectively clearing strips of the habitat. Even though the area of habitat loss can be small, the fact that you now have a number of very small isolated 'islands' means that, because of the species–area relationship (remember Chapter 3 and the relationship between the number of species in an area and the size of that area), each section can only support a very small population. There are three issues with this. First, any negative chance events, such as disease or environmental change, will make such small populations more vulnerable to extinction than they would a larger population inhabiting the same overall habitat area. Second, if any of the species present, for whatever reason, actually needs a large habitat area, they are in trouble. And finally, we know that small areas of habitat are more influenced by their surroundings than large-area habitats, much in the same way as it would take an ice cube one cubic metre in size much longer to melt than it would one thousand separate ice cubes made from the big one. So how the creatures in the small bit interact with their wider environment may be completely different from the way they would if they lived in a larger area.

One of the practical implications of the species–area relationship is that you can also work out how many species you will lose if you reduce their habitat size by a known amount. This is related to, although greater in its scope than, the idea of fragmentation we've just considered. Thus, habitat loss has been estimated to affect nine out of ten threatened birds and plants, and four of five threatened mammals. Specifically, we know that more than 300 bird species in Asia are threatened as a result of development, farming and logging. Information for marine life is understandably more difficult to acquire. Half of the live coral cover on coral reefs has been lost over the past 150 years. Currently, about three quarters of the remaining coral reefs are considered as threatened, although some progress has been made in conserving them. Since 1980, about one third of all mangroves have been lost. In 1996 Brooks and Balmford reported the case of the Atlantic forests of South America where 90% of the forest had been destroyed but no species, or at least no bird species, had yet become extinct. Basically, because of the species–area relationship, we know that the remaining 10% of forest is not enough to sustain all the species that were there. In the case of the birds, and indeed all of the species that are in the forest, many, many of them are effectively extinct. They just don't know it yet. This is referred to as 'extinction debt'.

One form of environmental degradation that has received a lot of investigation is pollution. The most serious threats come from chlorinated solvents like carbon tetrachloride, chlorinated aromatics like polychlorinated biphenyls or PCBs, components of petrol/gasoline, polynuclear aromatic hydrocarbons or PAHs, and trace metals such as zinc and cadmium. The input of chemical contaminants into the environment can potentially pose a threat, either directly or indirectly, to species and whole ecosystems. Interestingly, pollution does not seem to be a common cause of extinction globally – but it may be locally. Take a really polluted river and, instead of the numerous snail and insect and crustacean species that characterise clean sites, there's often just the one type of measly worm (sorry, measly worm). A relative of the earthworm, and a good, abundant food for any fish that can survive the filth, these measly worms belong to one of the 'survive anywhere' groups. There's still life, and it may be in abundance, but it's a very poor substitute for what was there.

Within the last few years, plastic pollution has been touted by some as one of the most pressing global issues of our time. Half of all plastics ever made were made in the last fifteen years. Eight million tonnes of plastic now find their way into the ocean each year, some of it estimated to take more than four hundred years to break down. There have been some biological effects of plastics on marine animals documented. Currently this is mainly due to entanglement or blocking of the gut, resulting in starvation. Without a doubt, our plastic crisis has highlighted, in the most visual way possible, our inability or unwillingness to deal with plastic pollution specifically, but also disposable products and even global pollution and waste more generally.

2) Direct exploitation

It's become a cliché that the Chinese word for 'danger' is also the word for 'opportunity'. But taking this 'both sides of the coin' approach seems appropriate when we probe the direct role of biodiversity 'resources' in consumption or production. In other words, treating biodiversity as a marketable commodity. In this chapter we're dealing mainly with threats to biodiversity – and direct exploitation and over-exploitation are definitely threats. They are, if you think about it, a subsection of habitat loss and degradation. But examining over-exploitation as a threat also doubles as an introduction to the use value, the economic value of biodiversity, which rightly should be included in the chapter that follows this one. Use value and over-exploitation are two sides of the same coin. Which side of the coin you give attention to depends very much on who you are and what you're doing, or want to do. As Kevin Gaston and I once wrote in the first edition of our *Biodiversity* textbook, 'What to one person is the "legitimate exploitation of natural resources" can be to another "the rape of the natural world".' So we will cover use of and threats to biodiversity together. However, this is no trivial task.

Home economics

One of the 'easiest' ways into this whole area is to ask the question, 'How much biodiversity have I used today?' In my case let's start from something as mundane as my waking moment on the morning of the day that I wrote this section. I woke with the sound of my bedside alarm in my ear. The alarm is made from plastic, which is derived from oil products, which themselves were once part of ancient biodiversity. The electricity powering my alarm, I think, has been generated from the burning of fossil fuels (overexploited). I pulled back my feather-filled (bird – no idea where from, overexploited or not), cotton-covered (plant – I don't know if it's overexploited or not) quilt, and lifted my head slowly off the cotton-covered (plant), synthetic-material-filled (fossil fuels again) pillow. I took a deep breath of the fresh but damp air coming in through my open window, air containing life-giving (and life given, as it is produced by plants) oxygen, which fortunately enough I do not have to pay for (yet). Once up I put the kettle on (electricity from fossil fuels), put a tea-bag (plant – fair trade but I don't know if the crop, a renewable resource, is overexploited or not) into my mug, added some milk (renewable resource) and pulled down the cardboard box (trees) of Shredded Wheat breakfast cereal (plant), placed some in a bowl, added some sliced bananas (plant) and raisins (plant – this is getting tiresome), and reached for the milk carton (cardboard – tree; covered with plastic – fossil fuel) wishing someone would make me kippers (overexploited) instead. All of this before the kids get up or I've gone for a pee.

This catalogue of domestic trivia makes an important point. If it is difficult, perhaps near on impossible, to get to grips with how much biodiversity one academic uses between getting up in the morning and eating his Shredded Wheat, how much more difficult is it to begin to comprehend the scale of our direct use of biodiversity globally. It is simply breathtaking and bewildering at the same time. Direct use is complex and multifaceted. It is often difficult, if not impossible, to tease apart those different facets. And at times we simply do not have the information that would allow us to know where the ever-shifting line between exploitation and over-exploitation lies.

Food, glorious food

Biodiversity, powered by the conversion of sunlight into energy-containing material (food, fuels and building materials), is fundamental to the whole of human history; and so we start with food. Biodiversity is the basis for all food industries both directly and indirectly. It is the basis of all the service industries too. The production and consumption of vegetables, fruits, nuts, meats, colourings and flavourings is essential for human existence and therefore, not surprisingly, is also very big business.

Much has been made in some textbooks of the (bio)diversity of foodstuffs. For example, someone has reported that up to 500 food-plant species can be found in home gardens of one village in Java. And about 50 different types of acacia seed are used as food by Australian aboriginals. In reality, though, of the 12,500 plant species considered edible, only about 200 have been domesticated and of those three quarters of our total food supply come from just a dozen different kinds: bananas/plantains, beans, cassava, maize, millet, potatoes, rice, sorghum, soybean, sugar cane, sweet potato and wheat. Global agriculture is responsible for more than nine tenths of our protein intake. There may not be much variety in the plants contributing to our food needs but the scale of exploitation is enormous, at around 3 billion tonnes produced every year.

Fungi are used extensively as foodstuffs with 350 species collected and eaten. Edible mushrooms are estimated to be worth US$42 billion per year. The production of many alcoholic drinks is underpinned by yeast.

As with plants, the number of animals used for food is restricted to a very small proportion of the species that could be used. Much of our use is culturally determined. Vertebrates, that is fish, amphibians, reptiles, birds and mammals, are high up on the list, as are substances derived from them, like cow's and goat's milk, and cheese. To a lesser extent, and depending on where you live, there are also molluscs (such as clams, mussels, snails, squid and octopus), crustaceans (such as lobsters, shrimps, prawns and crabs), insects (beetles, moths and the produce of bees, honey) and even echinoderms, some sea cucumbers and a few sea urchin species. Average global meat production in 2013 was 320 million tonnes, four times what it was fifty years ago, and yet it is still about an order of magnitude less than edible plant production.

The global seafood catch was 96.4 million tonnes in 2018 (a 14% increase over the period 1990–2018, and worth, in total, around US$100 million). Interestingly, for the first time wild catch was outstripped (just) by global aquaculture (114.5 million tonnes, an increase of 527% from 1990 to 2018, with a farm-gate sale value of US$263.6 billion); 88% of the catch was for human consumption. There are, however, still more people employed in fisheries (39 million) than in aquaculture (20.5 million). The most popular fish species is the anchoveta, weighing in at 7 million tonnes annually, followed by Alaskan pollock (3.4 million tonnes) and skipjack tuna (3.2 million tonnes). Fish currently provide 3.3 billion people with one fifth of their animal protein.

The other side of the coin is that many of the uses of biodiversity for food are unsustainable, at least in their present form. The history of hunting is such that often species have been pursued right up to the last viable population, or even the last individual. This is, for instance, the story of the whaling industry. Not surprisingly then, as we've already seen, hunting is the third most common reason for recorded extinctions (although not just for food but also for clothes and other commodities). One of the best examples of unsustainable practice is the hunting for bushmeat in some tropical forests. The Amazon Basin produces 1.8 million tonnes of mammal meat annually. Of this, only 150,000 tonnes are hunted. This means that you only need around a 15% increase in the rate of production to make up for the bushmeat taken as food. Contrast this with the Congo Basin. Here the production rate is just slightly greater than in the Amazon at 2.1 million tonnes of mammal meat annually. Currently 4.9 million tonnes are hunted every year. You don't need to be a mathematician to see that this is completely unsustainable. In 2020, *The State of the World's Forests*, a report from the Food and Agricultural Organisation of the United Nations (FAO), noted that 1 billion people depend on 'wild' food globally.

Another high-profile example of how our use of biodiversity for food can be unsustainable is commercial fishing. The middle of the last century saw a dramatic increase in global fishing efforts. It is claimed that the first major, well-documented fisheries collapse was the Peruvian anchoveta in the early 1970s, but there are a large number of earlier cases which often had profound effects locally. I even remember, as a child in Glasgow in the 1960s, my mum chatting with other mums about the disappearance of the herring (again) from the shops and from the local seas. The percentage of fish stocks that are fished sustainably has decreased from 90% in 1974 to 66% in 2017. The ocean's large predators, the sharks, the tuna, and the swordfish, are now an order of magnitude less abundant than they were.

However, apart from its obvious impact on the commercial species of interest, fishing activities have a much wider impact on biodiversity. Massive reduction in one species invariably has knock-on effects for others, often many others. For example, over-exploitation of herring and capelin in the north-eastern Atlantic led to the collapse of many large fish, seabird and sea mammal populations.

Also, the methods used to catch fish are rarely selective. Often many tens or even hundreds of different species – fish and invertebrates – are caught 'accidentally'. I have some photographs that I took of all the animals present in a bottom trawl deployed to catch scampi (Norwegian lobster, *Nephrops*). Scampi are bright orange-red, so relatively easy to see. But in a half-metre-high pile consisting of many, many hundreds of sea urchins, starfish, swimming crabs, dead man's fingers, clams and brittlestars (indeed almost every animal phylum is represented) there were only nine scampi. A number of other fishing methods can result in the capture and subsequent death of even larger creatures such as turtles and dolphins. Overall, such bycatch can be substantial – in the order of a quarter of a million tonnes annually. This is to say nothing of the habitat destruction caused by such methods as bottom trawling, and the environmental effect of lost nets and lines and the like. There are some areas in the Irish Sea, the North Sea and the English Channel where the scouring and subsequent rescouring of the seabed is one of the persistent features of the landscape.

Ultimately, many of our current food production and consumption practices are unsustainable. Furthermore, the 2017 and 2018 editions of the FAO-published report *The State of Food Security and Nutrition in the World* highlighted that conflict and climate change were undermining food security, with the 2019 edition showing that the situation was being exacerbated by economic slowdowns and downturns. In 2019 an estimated 2 billion people globally did not have regular access to safe, nutritious and sufficient food. Now, in the 2020 edition, the COVID-19 pandemic and an unprecedented desert locust infestation in East Africa are highlighted as having even further exacerbated the situation in ways that no one could have anticipated. The only thing that is predictable is that global food demand will increase substantially.

Industrial materials

An incredibly wide assortment of industrial materials is provided by biodiversity. They range from building materials like wood or grasses through to glues, dyes and even rubber and perfume. The scope of this facet of exploitation is so vast it is almost impossible to get a handle on. To make the point, let's focus in on one aspect where comparatively good information is available. Every year 3.8 billion cubic metres of wood are harvested, with worldwide exports worth over US$6 billion. While our source of timber is based on 1,575 species contained in 103 families, one third of that wood comes from trees belonging to just three of those families. But wood is not only used as an industrial material. Two billion people – almost one third of all humankind, and its poorest members at that – are reliant on fuelwood and fuelwood alone as a source of energy for cooking and keeping warm. There are currently an estimated 2.5 billion cubic metres of fuelwood or equivalent. Even allowing for many hundreds of schemes to increase the amount of fuelwood available (tree planting, for example), the demand in 2020 is estimated to be between 2.4 and 4.3 billion cubic metres. This is just one aspect of exploitation and over-exploitation of industrial materials, but the figures speak for themselves.

Having been so focused on what is a relatively obvious material to look at, wood, let's now take the Fungi, a grouping one might not have thought would be associated with many industrial materials. Here is a short eclectic mix of 'why, I did not know that' and 'biodiversity as a source of industrial' materials facts, focused on fungi. Enzymes from fungal species are used to degrease leather hides, accelerate the pulping process in the production of paper, and break down excess bleach used in cotton processing. Fungal products are now used as replacements for polystyrene foam, numerous building materials, and in the construction of synthetic rubber, plastic car parts and, who could have guessed, leather. The icing on the cake, I suggest, is the production of Lego™ pieces using itaconic acid extracted from a fungal species called *Aspergillus*.

Now for something related, but a little different. There are *industrial materials* that are, in one sense, one step removed from biodiversity itself, but which wouldn't exist without it; here we're speaking of the idea that by imitating biodiversity we can design better products. Let's choose two from so many examples of such biomimicry. First, termites have built air exchange systems to ventilate their tall mounds. Their design has inspired and informed the design of climate control in what are termed biomimetic buildings, like the Eastgate Centre in Harare, Zimbabwe. Similarly, the 180-metre-tall tower in the City of London called the Gherkin has a ventilation system similar to that of sponges, designed so that air flows through the entire building. Second, in 1941 entrepreneur and engineer George de Mestral was out for a walk with his dog when he noticed that burrs had attached to both of them. Intrigued, he examined the burr under a microscope and saw small hooks on it which were firmly but not irreversibly attached to loops in the fur/fabric. He came up with the idea of mimicking this structure as a fastener. The design he subsequently marketed was Velcro®, made up by combining the French word for 'loop' (*velours*) with the French word for 'hook' (*crochet*). The *Global Velcro Hoop & Loop Market Report 2019–2024* estimated that Velcro® was worth US$2.02 billion in 2019.

When it comes to the multifarious industrial uses of biodiversity, like 'the other things Jesus did', 'if all were written down, the world itself, I suppose, would not hold all the books that would have to be written.'

Medicine sans frontiers

Biodiversity provides many of the world's medicines and drugs and has done for millennia. In 2017 it was estimated that at least 38,187 plant species were used in medicine. One of the earliest must be the opium poppy, as one of its chemical constituents is the powerful painkiller morphine. As four out of five people in the world do not have access to modern medicine, natural products from medicinal plants play a key role in the health of most of the world population. About 20,000 plants are used in traditional medicines. The annual global market for herbal medicines is estimated to be worth US$60 billion, with the global industry based on traditional Chinese medicine worth US$83 billion.

Many of the commercially available medicines have their origins in biodiversity. Aspirin is derived from acetylsalicylic acid present in the meadowsweet plant. Of all the new drugs approved from 1940 to the end of 2014, half of them were made from natural products or their derivatives. And 35% of all medicines are derived from natural products, and are worth US$385 billion per year globally (2019 figures). The bestselling and most popular anti-cancer drug ever (which peaked at US$1.6 billion a year in 2000) contains paclitaxel (brand name Taxol™), which is used to treat breast and ovarian cancers. It was first isolated in 1971 from the Pacific yew tree, a species which used to be discarded as worthless during logging operations. The hit rate for producing new drugs is better in nature than it is in the laboratory. One in 125 plant species produced a major drug compared with one in 10,000 for artificially constructed chemicals. And don't forget the Fungi. The antibiotic penicillin comes from the fungus *Penicillium*; vitamin B2 used in vitamin supplements is produced by fermentation of a fungus; and 15% of all vaccines are made in yeast.

Even some of the more obscure animal groups that barely got a mention in Chapter 2 can turn out to be important in producing new drugs. For example, drugs for the treatment of cancer have been isolated from sponges (phylum Porifera), sea mats (phylum Bryozoa) and sea squirts (phylum Chordata). It has been estimated that 50% of all 'substances' isolated from marine animals (e.g. shark's liver) and plants since the early 1970s have anti-cancer properties.

The other side of the coin is that there is the possibility of over-harvested medicinal plants or animals becoming endangered or even extinct. One of the earliest examples of over-exploitation must be the herb silphion, which once was common in northern Libya. It was used, with some effect, as an oral contraceptive, albeit a fairly pricey one. It was so popular that it was effectively extinct by the third century CE. Today even bits of currently very rare animals, such as turtles, tigers and rhinos, are in great demand for some traditional medicines, often irrespective of whether they possess any medicinal qualities. It is extremely difficult to estimate the global threat to biodiversity of collecting animals and plants for medicinal use, as most extinctions will probably occur on a greater scale through habitat destruction and degradation.

Ecotourism

Travelling to locations to see and experience wildlife is known as ecotourism. Not only does it use biodiversity as a resource but it's founded on biodiversity by definition. Ecotourism is a relatively modern phenomenon and is also big business. Global ecotourism generates US$77 billion (2009), which is 5–7% of the total travel and tourist market, and it was still growing, at least before the beginning of 2020. Protected areas worldwide received 8 billion visits per year generating US$600 billion direct in-country spend and US$250 billion consumer spend. It has been estimated that a single African elephant could generate US$22,966 for the tourist trade every year, and because elephants can live for many decades, each elephant could generation US$1.6 million during its lifetime. Also, you don't have to travel very far to qualify as an ecotourist. According to the World Association of Zoos and Aquariums (WAZA), every year there are over 200 million visits globally, with three out of every five visitors aged eleven or younger.

While it is true that ecotourism, by raising awareness of the existence of, and threats to, biodiversity is potentially a very positive development, there could be some significant downsides. Responsible ecotourism could deliver real environmental and social benefits, but it may also be taken merely as a new marketing opportunity – what has been christened 'green greed'. Even with the best motives in the world, ecotourism may have a negative impact on the wildlife or environment at its centre by increasing pollution and travel to and from the destination, and degrading the environment as facilities need to be built to house and entertain ecotourists. Sheer numbers may well overwhelm the destination and exceed the carrying capacity of the area. Many reading this chapter will be keen to know, given everything said up to now, whether ecotourism helps or harms the environment. The answer is undoubtedly both. What is not clear is which outcome is most prevalent.

Controlling the natural world

Use of natural enemies to control problem species is a common feature of biological control. Leafy spurge is a weed introduced into the US in 1827. It displaces native plants and restricts cattle grazing. Biological control of this plant began in the 1960s and has entailed the introduction of 15 foreign (non-indigenous) insect species. The most effective have been the flea beetles which destroy 80–90% of the plants in an area. The economic benefits have not been fully assessed, although one study has suggested a benefit–cost ratio of between 56:1 and 8.6:1 in the US and Canada, respectively.

A more detailed study of the economics of control of the red waterfern has been made. This waterfern is native to South America but it invaded waterways in South Africa, causing economic loss to water users, mainly farmers, but also those using waterways for recreation. The frond-feeding weevil was released as a biological control measure in 1997, and within three years the waterfern it fed upon was largely under control. The economic cost of the 'damage' caused by the invasion was US$589 per hectare per year. The cost of biological control was US$278 per hectare per year with an initial one-off investment cost of US$7,700. So in 2000 the benefit–cost ratio was estimated at 2.5:1, but by 2005 had increased to 13:1 because the treatment had been so successful. Some other success stories include the control of the cassava mealybug in Africa by an introduced wasp species, with annual savings in excess of US$250 million, and control of the banana skipper butterfly in Papua New Guinea, with annual savings of around AU$201 million.

In summary, when it is successful, the economic gains of biological control can be huge, but when it goes wrong, it really does go wrong. This is intimately linked to the introduction of species to new areas, which we will come on to.

3) Climate change

> The furnaces of the world are now burning 2,000,000,000 tons of coal a year. When this is burned, uniting with oxygen, it adds about 7,000,000,000 tons of carbon dioxide to the atmosphere yearly. This tends to make the air a more effective blanket for the earth and raise its temperature. The effect may be considerable in a few centuries.
>
> *Waitemata & Kaipara Gazette*, 14 August 1912

One of the most serious indirect effects of pollution is climate change. By burning fossil fuels, we release large amounts of carbon dioxide into our atmosphere. Carbon dioxide is a greenhouse gas. It helps to trap energy from the sun's light in our atmosphere, causing the air to warm up. The ocean absorbs a large proportion of the carbon dioxide, ameliorating the warming that should take place in the atmosphere. However, this amelioration comes at a cost as the carbonate chemistry of the seawater changes in quite complex ways. Amongst other things this results in a reduction in the alkalinity of the water (termed 'ocean acidification') and a change in the availability of substances used by marine animals to build calcium carbonate structures, such as the shells of snails and mussels and the outer covering of shrimps, lobsters and crabs.

There have been a large number of *natural* climate changes in the past (see Chapter 4), but nothing like on the timescale we're experiencing today. From pre-industrial times to 2017 we have been responsible for a warming of about 1°C on our planet. This has been most marked over the last 30 years where average temperature has risen by 0.2°C every ten years. The Intergovernmental Panel on Climate Change (IPCC) have warned that a 1.5°C increase in global average surface temperature is a limit beyond which climate change will have 'devastating consequences'.

We've already seen that where species live (their biogeography) is greatly influenced by climate. Rapid global climate change will alter, and is altering, species distributions. At best, climate change may merely shift these distributions. Tropical species will head into more temperate regions, and temperate species to more polar regions. The main losers are likely to be the species that prefer the coolest habitats. They will find themselves, quite literally, with nowhere to go. But that's at best. It is extremely probable that many species will not be able to either adapt or migrate as rapidly as the world is warming up. And even if migration were a possibility, many of the barriers to movement associated with our presence – roads, cities and the like – may deprive species of the opportunity. Global agricultural production will be affected, but regional differences make it difficult to say what the overall outcome will be. There will be physical habitat loss as sea levels rise and many low-lying areas, including marshlands, will be flooded. Some authorities have estimated that one quarter of all species could become extinct directly as a consequence of global climate change. The bottom line is that we really don't know. But this is not an ignorance to take comfort from. Given the seriousness of global climate change for biodiversity, and for so many aspects of human endeavour and existence, it is disappointing that many countries, or at least their leaders, still do not take the threat seriously.

4) Introduced species

Nikolas Kaplanis and colleagues from Scripps Institution of Oceanography discovered a small patch of an invasive seaweed, *S. honeri,* at Bird Rock in the summer of 2013. By the winter of that year it covered the majority of the cove. This is just one of at least twenty-seven non-native seaweed species that have been found in California over the last forty years.

After habitat loss, direct exploitation, and how these threats are amplified by climate change, the biggest threat to local and global biodiversity is thought to be invasion by non-native species. The invaders, as we've already seen, are sometimes introduced deliberately for the purposes of biological control. Others are introduced both intentionally and unintentionally, mainly through trade and tourism. Approximately 400,000 species have been introduced. Only one in ten of these have become successfully established. Of those established species, only a further one in ten go on to become pests. Not a great success rate for the making of a pest, but still a very small number of pests can play havoc with their new environment and its inhabitants.

It is important to realise that invasions have always happened. What is critical now is the large increase in their number as a result of increased human activity. For example, a colleague of mine, when I was lecturing at the University of Sheffield, studied the many types of seeds present in the mud on the wheels of lorries coming into Plymouth from France and Spain. There is also work that shows that the number of invaders in nature reserves is directly related to the number of human visitors coming into the reserve.

Cumulative records of invasive species have increased by about 40% since 1980. The cost of damage due to invasive species globally was estimated at US$1.4 trillion, 5% of the global economy, in 2001. Half of all the threatened species in the US are thought to be at risk because of invaders. The invaders currently in the US are thought to cost about US$137 billion annually, both in economic damage and in trying to control them. Invasive plant species already cover 400,000 square kilometres of the country. They continue to spread at a rate of about 12,000 square kilometres a year. The Galapagos Islands have almost as many introduced species as native ones. The Baltic Sea now has an additional 100 new invasive species, a third of them native to the US Great Lakes. This said, a third of the 170 aliens in the Great Lakes are originally from the Baltic.

Pests can play a major role in biodiversity decline. These invaders either eat the threatened species (predators) or live off them (parasites). The African Great Lakes of Malawi, Tanganyika and Victoria were famous for the large number of endemic species of cichlid fishes they supported. However, someone had the great idea of introducing a really big fish-eating predator, the Nile perch, into these waters because it would improve the fishing. The Nile perch has become established in Lake Victoria and is quite simply eating to extinction many of the cichlids.

I accept that it is easy to cast a large predator in the role of 'baddy'. However, even cute furry creatures can ravage whole ecosystems in a less than cute and cuddly way. Take the European rabbit, for example. It originally occurred on the Iberian Peninsula, south-west Europe. However, the ancient Romans introduced it to Italy not because it was cuddly but because it was tasty. Rabbits, at least those that weren't eaten, did well in Italy and in other European countries. They came to England in 1066 along with the Norman Conquest and thereafter did a bit of island hopping during the Age of Exploration. The classic *thing-not-to-do* was done by Thomas Austin in 1859. He was an English landowner who had emigrated to Australia. I remember missing the BBC news and brown HP sauce when we lived in Canada; Thomas specifically missed shooting rabbits for fun. So he imported twenty-four individual rabbits to Victoria, from England.

Today, rabbits are one of Australia's greatest environmental problems. They are the main cause of habitat destruction and degradation. They devour crops. Sixteen rabbits eat as much as one sheep. In a *good* year there can be a billion rabbits. They alter whole ecosystems just by how they feed. They threaten elements of the native fauna with extinction, in particular marsupials such as the greater bilby (now endangered) and the burrowing bettong (now extinct on the mainland). They directly compete for food and habitat, ousting many, like the rufous hare-wallaby, from their burrows. The view today is that the rabbit population is under control, mainly because of the *help* of two introduced viruses, but as one expert has said, 'The situation...appears precarious and requires continuous surveillance.' Biological control of the rabbits costs AU$20 million annually but has produced a benefit of AU$70 billion (2011) for agriculture over the past 60 years.

A good example of where we have a successful invasion but the invader turns out to be benign is the terrestrial 'shrimp' originating from the southern hemisphere. Commonly referred to as 'landhoppers', these shrimps have successfully invaded leaf-litter habitats both in Europe and in the US. One species I have worked with is the Australian *Arcitalitrus sylvaticus*, which has been established in, amongst others, a number of areas in California, with the earliest record dating from 1967. I've seen and identified these landhoppers from in and around San Diego, including Balboa Park, beside San Diego Zoo. This species can occur in densities of thousands per square metre but they do not seem to have any negative effect on the native fauna, except for being a nuisance by invading our house when it rains a lot. We thought maybe they competed with woodlice, but we could find no evidence that they did. In the UK we compared the identity and abundance of animals that lived in areas where landhoppers (a related species *Arcitalitrus dorrieni*) were common and areas where they were absent. There did not seem to be any difference. Having studied them since 1986, we have uncovered no negative effects of landhoppers on natives. And that seems to be the case for nine out of ten successful invaders.

The domino effect: Extinction cascades

It is not uncommon that the extinction of one species leads to the extinction of one or more different species. This is called an extinction cascade. For example, in New Zealand hundreds of years ago, giant eagles fed on a flightless bird species, the moa. The Maori hunted the moa to extinction. With its main food item off the menu, the New Zealand giant eagles became extinct. Another example, but this time of a species that is right on the brink of extinction, is the *Calvaria* tree on the island of Mauritius. Passing through the gut of a flightless bird, the dodo, was the key to the germination of the seeds of this plant. The dodo was hunted to extinction in the late seventeenth century. Since then *Calvaria* seeds can still be found but they cannot germinate. The species is now hanging on, just, in the form of a few very old trees.

Lian Pin Koh and his colleagues modelled co-extinction (the loss of a species upon the loss of another) for a set of interrelated species including wasps, parasites and their hosts, butterflies and the plants that fed their caterpillars, and ants. They estimated that for every one species on the endangered list, there are 6,300 related species that are *co-endangered*. If they're right, we will have to revise the numbers of endangered species that we discussed in Chapter 4.

Not all extinction cascades are global. There are also local examples where the extinction of species in one area has dramatic knock-on effects for all the other species in that area. Linked with the threat to biodiversity by invaders that we've just dealt with, there is the story of the mysid shrimp (sometimes called opossum shrimp), which was introduced into Flathead Lake, Montana, US.

The shrimp was supposed to be an extra food source for the economically important salmon in the lake. Unfortunately, the shrimp ate very large numbers of the native microcrustaceans. The microcrustaceans were the mainstay of the salmon, so the salmon numbers collapsed to such an extent that populations of bald eagles, brown bears and human residents that relied on the salmon were all adversely affected.

Some light relief: Complete elimination of biodiversity by extraterrestrial means

'You cannot help but get a big flash when objects meet at 23,000 miles per hour,' said Dr Pete Schultz when at 1:52 A.M. on 4 July 2005 an impactor released by the NASA spaceship *Deep Impact* slammed into Comet Tempel 1. 'The heat produced by impact was at least several thousand degrees Kelvin and at that extreme temperature just about any material begins to glow.' The *Deep Impact* mission was all about investigating the origin of the solar system, studying the make-up of the comet by throwing things at it. However, it also illustrated very graphically how something the size of a refrigerator hitting an object floating in space can cause an awful lot of damage.

Consider then that there are more than a thousand objects, asteroids or comets greater than 1 km in diameter, and more than a million greater than 40 m in diameter (modest office block size), which regularly come close to Earth, and could possibly strike it. These Near Earth Objects (NEOs), if they cross our path and get through our atmosphere, could cause widespread environmental damage and biodiversity loss on a local (for an NEO 40 m to 1 km in diameter) or even global (for an NEO 225 km in diameter) scale. None of the currently recognised NEOs (which can be found updated daily on a NASA website) are thought to be a threat, but how do you predict the behaviour of a presently unknown object? At the end of April 2020 a 'potentially hazardous' asteroid, 2 km in diameter, known as 1998 OR2, came within 4 million miles of Earth. Interestingly (bizarrely) the observatory that was tracking it tweeted, 'It even looks as if it's wearing its own mask.' And then a few months later in August 2020, Asteroid 2020 QG (merely the size of an SUV), moving at 12.3 km per second, became the closest asteroid observed to pass the Earth – a mere 2,950 km away, flirting at the same sort of altitude occupied by many of our telecommunications satellites.

On average, a 'dangerous' NEO (2 km in diameter, 1 million megatons of energy) does collide with Earth once or twice every million years. Currently, you and other bits of biodiversity have a one (or less) chance in a hundred million of being killed by an asteroid collision. But don't let it keep you awake at night. Despite a small army of dedicated asteroid hunters worldwide, and the NASA-backed Spaceguard programme looking out for NEOs, your first indication of an impending impact is likely to be a spectacular, short-lived (for you at least) flash and shaking, not at all like the Hollywood movies *Deep Impact* and *Meteor*.

In 1980 the physicist Luis Alvarez and his son Walter, a geologist, proposed that an asteroid 6–15 km in diameter (100 million megatons energy) collided with the Earth about 66 million years ago (coinciding with the end of the Cretaceous period). This resulted in the Chicxulub crater on Mexico's Yucatán Peninsula. The impact would have penetrated the planet's outer covering, much as we saw with the *Deep Impact* experiment, scattering dust and debris into the atmosphere, and causing all sorts of natural disasters. Again, as in the *Deep Impact* experiment, the heat produced would have incinerated all life in its path. So death by asteroid, both directly and indirectly, is seen as the most likely cause of the extinction at the Cretaceous/Tertiary boundary.

In an article in the magazine *Science* from May 1991, evidence was presented for the rapid demise of most life forms (over a period of 10,000 years rather than the 10 million that used to be proposed) in a mass extinction event at the end of the Permian period. It has been suggested that the late Permian, late Cretaceous and many other mass extinctions could have had an extraterrestrial cause, with the slate of life almost being wiped clean every 100 million years or so. Palaeontologist David Raup, of the University of Chicago, suggested that roughly 60% of all species extinctions may have been caused by impacts.

So extraterrestrial objects hitting the Earth has been in the distant past, and continues to be to the present day, a remote but serious threat. It is entirely possible that much of the history of biodiversity has been shaped to some extent by such collisions. The extent to which biodiversity will be 'modified' by such impacts in the future is unknowable. And this is the one threat to biodiversity where, if the light around us gets suddenly brighter, we can rest assured that it's genuinely not our fault.

The ultimate cause of biodiversity loss: You and me

Once upon a time there were two people... now look how many

For most of the 3 million plus years of our existence, humans have lived short and often precarious lives as ad hoc exploiters of biodiversity – what we call hunters and gatherers. However, as we've already seen, even a prehistoric world population of around 10 million people could have had a major negative impact on biodiversity. A more organised approach to exploiting biodiversity began to emerge 6,000 years before the present, at first in the Middle East – we call this agriculture. This new lifestyle resulted not just in different kinds of communities, but larger-sized ones. By the first century CE the world population was estimated to be 300 million, growing steadily to 760 million by 1750. The advent of the Industrial Revolution resulted in an increase in living standards for some at least, and coincided with a respite from the more severe famines and epidemics that had until then plagued communities. Thus, by 1800 the world population had reached the one billion mark, two thirds of whom resided in Asia, and a further one fifth of whom lived in Europe. Although it took a few million years to reach the first billion, the second billion took only 150 years. Between 1960, the year I was born, and when I turned fifteen, another billion were added! This increase was to some extent fuelled by global attempts to reduce infant mortality (e.g. introduction of immunisation programmes against diphtheria, cholera and other life-threatening diseases) and enhanced global food production as a result of use of fertiliser, more efficient farming practices and the introduction of new disease-resistant crops (e.g. new strains of rice). The twentieth century started with 1.6 billion souls on the planet and finished with 6.1 billion. At the exact time I write (10.50 GMT, 5 August 2020) there are 7,802,970,469 people (according to Worldometer). At 10.51, exactly one minute later, the total is now 7,802,970,628. The human population has multiplied more rapidly than before and this exponential growth seems set to continue (although growth rates are showing some reduction due to declining birth rates) (Fig. 22). Right now, three children are born every second. Each day a quarter of a million children are born. We saw earlier that 10 million of our ancestors had a profound and irreversible effect on biodiversity; consider now that the total number of all those ancestors represents a mere forty days' worth of births at current rates.

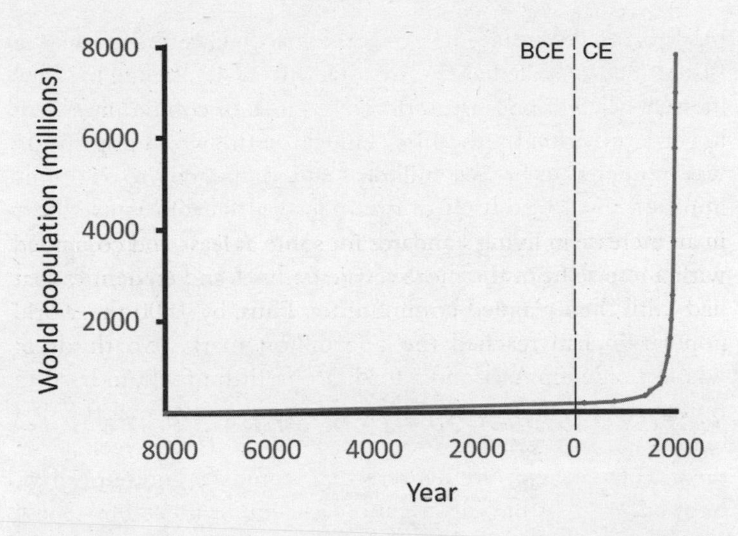

Figure 22 World population growth from 8000 BCE to the present (2020). Raw data obtained from Worldometer (www.worldometers.info/world-population).

Not just population size but where people live

Most of this growth is forecast to occur in lower-income countries in Africa, Asia and Latin America. Africa in particular will see huge increases in population, although many of the current predictions do not always take into account the impact of the HIV/AIDS tragedy, nor indeed have they factored in the consequences of COVID-19.

Overall food production is still sufficient to meet global demand. That said, many regions of the globe cannot produce enough food to support their populations. One place where this is glaringly obvious is equatorial Africa. The whole area is fast being transformed into a desert. The fast-growing population is rapidly clearing the forest for agriculture and for fuelwood.

About a century ago, half of Ethiopia was forest. Today, only about 11.2% is forested. Sub-Saharan Africa has the highest birth rate and shows the greatest rate of population increase anywhere in the world. The population doubled between 2006 and 2016, but food production did not. Such rapid population growth cannot but lead to greatly increased rates of habitat loss and fragmentation. However, we would also do well to remember that at the moment the most pronounced pollution and energy-hungry 'hotspots' are countries in Europe and Asia.

Not just population size but what people do

While human population size is considered the foremost driver of biodiversity loss, little attention has been paid to how this is related to the number of households that exist. More households means greater consumption of wood for fuel, habitat alteration for home building, and increasing greenhouse gas emissions. Lui and his colleagues found that household numbers increased globally between 1985 and 2000, and particularly in those countries containing biodiversity hotspots. Even in cases where population size declined, the number of households still increased substantially. These investigators proposed that a reduction in average household size alone would add a projected 233 million additional households to hotspot countries from 2000 to 2015. They appear to have been broadly correct, although subsequent work has emphasised the complexity of the myriad demographic changes complicating the picture. But broadly speaking such proliferation in household numbers translates into urban sprawl, greater per capita resource consumption, and so is a further drain on biodiversity

Finally, it is not just total numbers, but the distribution of population and what that population does that matters. Migration into environmentally sensitive areas has a big impact, while concentration of human activities in towns and cities focuses pollution onto small areas. Energy consumption increased by at least three orders of magnitude from the time of the agricultural revolution up to the present day. Over the period 1990–2016 global increases in energy use jumped by almost 60% to a total of 568 exajoules (1 exajoule = 10^{18} joules; a joule is a unit of energy equivalent to 2.39 food calories). The population increased by just 29% over the same period. The scale of this consumption is difficult to comprehend and noticeably outstrips the rate of population growth. The growth in energy use during this period was driven mainly by Asia, where the Chinese total energy supply increased 400%, accounting for just over a fifth of global energy use. Humans use a quarter of total rainfall and about one half of all available fresh water running off the land, mainly for agriculture. Water withdrawals from lakes and rivers have doubled in the last forty years. The flow from many rivers has been reduced dramatically with, at times, water in the Yellow River in China and the Colorado River in North America (amongst others) never reaching the ocean. About 1.5 billion people live beside rivers that are drying up.

It is inconceivable that human impact on this scale would not have major detrimental impacts on biodiversity.

It's the poor that do the suffering

As well as the numbers of human beings increasing exponentially, the economic fate of these new additions is also changing both rapidly and dramatically, with major implications for biodiversity. In 1960, when I was born, one fifth of the world population had two thirds of the world's money. By the time I took up my postdoctoral fellowship in Canada in 1989, that two thirds had increased to over four fifths. During that same period, the poorest fifth of the world's population saw a reduction in their income from just over 2% to around 1% of global income. In 2002 it was estimated that the richest 1% earned the same amount as the poorest 57%. By 2020 the richest 1% had more than twice as much wealth as 6.9 billion people, most of the people on Earth. And, according to the charity Oxfam, that richest 1% 'avoid' as much as 30% of their tax liability, and only 4 cents in every dollar raised by taxes comes from taxes on wealth. Also according to Oxfam, between 2017 and 2018 a new billionaire was 'created' every two days, and the 26 richest of those billionaires (it was 61 in 2016) own as many assets as the poorest half of the global population. Over this same period it was estimated that the very richest saw a 12% increase in wealth while the very poorest experienced an 11% reduction. And in January 2020 a new Oxfam report, *Time to Care*, only served to highlight the scale and nature of the problems with global inequality. The world's 21,563 billionaires have more wealth than 60% of the global population. The report paints the picture of a wealthy elite accumulating their massive wealth at the expense of ordinary folk, and particularly girls and women – for example, they are estimated to provide 12.5 billion hours of unpaid care work daily; this is worth a minimum of US$10.8 trillion a year, three times the value of the tech industry globally.

And even more recently, Oxfam has warned that the global responses to COVID-19 could push half a billion people into poverty.

Almost half of the global population is living on less than US$5.50 a day. Contrast this with the money locked up in the world's tax havens, thought to be around US$36 trillion. More than enough to pay for UN Millennium Development Goals and eradicate poverty in Africa twelve times over.

The US, with one twentieth of the world's population, consumes about a quarter of the world's resources and generates about a quarter of the world's waste. Each US citizen consumes the same amount of energy as 2 French citizens, 16 Bangladeshi citizens and 3 Chinese citizens (even though we've seen that energy use by China ranks first in the world). Another way of thinking about the energy use of a US citizen is that it is equivalent to the energy released by 165,000 sticks of dynamite or 31,000 burritos (thankfully burritos release their energy more slowly than dynamite).

By and large, countries of the southern hemisphere tend to be poorer than those in the northern hemisphere. There is a latitudinal gradient in poverty that mirrors the latitudinal gradient in species richness we saw two chapters ago. The greatest poverty coincides with the greatest species richness. But, irrespective of where you are, the extent to which a country's earnings are skewed towards a small number of wealthy individuals (income equality) has increased over the years. Every way you look at it, the rich get richer while the poor get poorer.

While a rapidly growing population cannot but impact on biodiversity, it is equally true that it is what those people actually *do* that has the greatest influence. Inequalities in wealth distribution seem to result in a huge scale of exploitation and pollution by the rich, and perpetuation of poverty as the poor are forced to degrade their natural environment in order merely to subsist. Damage to ecosystems impacts most directly on the poor. It is the poor that suffer the greatest effects of polluted environments. It is the poor that suffer the loss of productive lands. It is the poor that suffer the loss of traditional sources of food, fodder, fuel and fibre when forests are destroyed. The poor suffer a greater vulnerability to local environmental degradation, which money to some extent alleviates. And it seems clear that such inequality is not inevitable – it's a political choice.

Recent advances in economics and in technology only seem to widen the economic gaps both within and between countries. According to a recent report by the World Bank, global wealth increased by two thirds from US$690 trillion in 1995 to US$1,143 trillion in 2014. While the share of that growth rose for middle-income countries (from 19% to 28%), there was next to no movement for low-income countries (which hovered around 1%, even though those countries' share of global population growth increased by 6–8%). If such growth is to be realised and we are to genuinely attempt to Make Poverty History with current lifestyles, we need access to the equivalent of three 'current Earths' with associated biodiversity. Without changing our mind on how we use and treat biodiversity, other threats – to economic growth, world peace, and quality of life issues – seem small fry compared with this, although each of them will be exacerbated by our current biodiversity crisis.

To conclude

The scale of our use and misuse of biodiversity is difficult to comprehend but is clearly exerting tremendous pressure on that biodiversity. Furthermore, the effect of declining biodiversity impacts the world's poor to a greater extent than the rich. The conservation and management of biodiversity in the short term is therefore arguably more critical for the poor than for the rich, because quite literally biodiversity use is more 'immediate' in their lives and livelihoods – it can be a matter of life and death. But even the rich cannot evade the serious consequences of rapid biodiversity decline. Faced with the complicated mixture of threats and challenges we've looked at in this chapter, with all their political, sociological, moral and biological implications, the question is exactly what *do* we and what *should* we value about biodiversity? This is the subject of the next chapter.

6

Are the most beautiful things the most useless?

> Remember that the most beautiful things in the world are the most useless; peacocks and lilies for instance.
>
> John Ruskin

> Anything that just costs money is cheap.
>
> John Steinbeck

'...and for everything else there's Mastercard'

I got engaged to my now wife, Fiona, in 1979. I bought her a beautiful (but modest – give me a break, I was still a student) gold ring. In return she presented me with a pure white Antoria acoustic guitar. The guitar is a tad battered and worn now. Its present market value is derisory. However, indirectly through the years it has more than paid for itself through gigs and functions, and it has brought (to me at least) inestimable pleasure. My white guitar is worth very little and at the same time it is priceless. Given how incredibly difficult it is to put a value on something as simple as an old guitar, how do we begin to ask the question, 'What is the value of all living things?' This is a particularly pressing question given, as we saw in the previous chapter, the enormity of the threats. So how do, or should, we value biodiversity?

As with my old guitar, it is impossible to satisfactorily disentangle all of the different types of value we could attach to biodiversity. Take the temperate rainforest on the west coast of Vancouver Island, Canada, where my family (and guitar!) spent some time in the late 1980s (Fig.1, p. 4). I was a postdoctoral fellow working at Bamfield Marine Station in the Pacific Rim National Park. Certainly, it was evident that real hard cash was generated from logging and associated activities, as well as fishing (big salmon!), surfing, kayaking, hunting, great walking and trekking (along the famous West Coast Trail, which ended at our door), tourism and ecotourism (lots of whale watching). But indirectly, and a little trickier to put a monetary value on, was the existence and well-being of the rainforest. For the forest was essential to protecting watersheds, keeping the soil in place and 'working' (with associated implications for water cycling and management), producing oxygen, 'mopping up' carbon dioxide, and many other free (to us) but essential biological 'services' – often termed ecosystem services. Even more difficult to quantify, and inextricably linked with all of the above, is the 'value' derived from the forest in terms of the cultures (indigenous and non-indigenous) it has helped to shape, the art, music and literary work, even the religious beliefs it has inspired. And how do you put a price on one of my own most incredible experiences – watching the sunset over the trees from the top of a hill close to Bamfield, and through those trees tracing the hundred or so Broken Group Islands bathed in shimmering gold out to the horizon?

In some ways it feels wrong, or inappropriate, to tease out all of these different interrelated values – even to want to do it in the first place. But against the backdrop of unprecedented biodiversity losses and the continuing pressure on biodiversity from the drive for profit, for livelihoods, for mere existence in some cases, we must try to get a handle on all of these different values, however unsatisfactory that might feel. The order in which we tackle the various values of biodiversity could be taken to indicate their value to the author. It doesn't. We'll use a generally agreed scheme devised in the mid-1990s for separating out the different values of biodiversity.

Costing a small planet

So in Figure 23 we have the value of biodiversity divided into a number of different categories. None of these are watertight, and in reality there is much blurring, and not just at the edges either. The total economic value of biodiversity can be divided into two largely self-explanatory categories – use value and non-use value. Use value can be further subdivided into utilitarian value (useful now), option value (possibly useful soon) and bequest value (useful to future generations – see p. 185). Utilitarian value can be further subdivided. The value of marketable commodities (production and consumption subject to

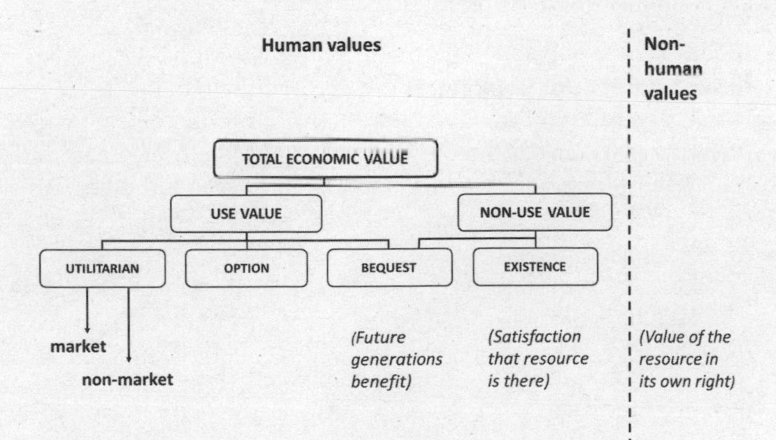

Figure 23 The value of biodiversity.

direct trading) is perhaps the easiest (!) to quantify. We've already covered this to some extent in the previous chapter while considering threats to biodiversity. The value of non-market commodities – aspects of biodiversity that are not directly subject to market trading, but could still in theory be given a market value, is where we'll start in this chapter. We'll attempt to put a price tag on non-market commodities (what Earth gives us, or does for us, seemingly for free). We'll then move on to look briefly at option and bequest values. To finish we'll discuss aspects of two sets of values that are often so intertwined that we may not be able to separate them. Existence value, 'satisfaction' that a resource is there, is one of the key non-use categories, and intrinsic value is the value of something in its own right independent of human values. We'll look at what different people and faiths have believed about biodiversity down through the ages as a way of attempting to get a handle on existence and intrinsic value as well as throwing some light on our overall approach to biodiversity.

Use now, pay when?

There are many things that biodiversity 'does', things essential for maintaining a planet that in itself is capable of sustaining life, and so keeping you and me alive. Making and maintaining our atmosphere, the recycling of nutrients, controlling species that could be pests, the acquisition of energy from the sun into our world through photosynthesis, the pollination of plants. These are a few of the services from nature that we perceive as getting for 'free' and which come under the umbrella-heading of non-market or indirect use value. Now, while it is true that many of nature's 'services' are not subject to direct buying and selling and so do not (yet) have a monetary value, we do not exactly get them for free.

What bees do for free is expensive

Take, for example, the indirect value of pollination, the reproduction method of many seed plants. This involves the transfer of genetic material between plants by insects and, in some cases, birds. Pollination is certainly valuable as without it many plants simply could not reproduce. No harvests, no food. But you couldn't put a monetary value on it. Or could you? In many areas of the world the numbers and types of wild bees are in decline. This has a knock-on effect in the form of reduced crop production. Reduction in pollination results in a reduction in the plant species that depend on these pollinators. So there is an economic cost to *not* having pollinators. But more than this, consider how costly it would be if we had to employ people to work in our fields manually exchanging pollen between plants. Take Sichuan province, China's 'world pear capital', where overuse of pesticides led to a dramatic reduction in bees, and so a notable reduction in pollination. In 2014 they employed workers to pollinate trees, paying each of them US$12–19 to pollinate 5–10 trees a day. They eventually found that it was cheaper to rent bees from elsewhere. In some US farms they use managed honeybees to pollinate crops to increase yields, while it is estimated that the wild honeybees still add at least US$1.5 billion to yields of the top six crops, indicating that crop production is still limited by a lack of pollinators. In 2008 an international group of scientists from France and Germany estimated that the service provided by pollinators could have a market value of US$217 billion, a figure which in the intervening years has jumped to between US$235 and 577 billion. All of a sudden what was of indirect value, and you seemingly got for free, presents you with a fairly hefty demand notice.

This idea of trying to cost nature's services is perhaps most tangible and hard-hitting in what has become a landmark scientific paper in our attempt to value biodiversity.

Costing the Earth – literally

In 1997 Robert Costanza and twelve colleagues suggested in the magazine *Nature* that it would be useful to attempt to put a monetary value on nature's services, the ecological systems critical to maintaining life on Earth, arguing that 'a number is better than no number.' Taken together these services are the basis for all human welfare. As such they must, in some way, be part of the total economic value of our planet. So Costanza and colleagues took published information, and made a few of their own calculations. They estimated the 1997 economic value of seventeen ecosystem services. Scaled up for the whole Earth, the total cost of those services came to around US$33 trillion per year. Not only was this, they claimed, a conservative estimate, but nearly all of that 'value' was outside the current market.

In 2014 Costanza and a different group published an update of the 1997 paper. The revised cost was now US$125–145 trillion per year, about four times the original conservative figure. What is more, they estimated that loss of ecosystem services, what biodiversity does *for nothing* and which, at the same time, contributes twice as much to our well-being as global GDP, was worth about US$20 trillion every year (1997–2011).

So assuming we actually could buy the things that the Earth does for us for free, could we afford them? In the Christian New Testament there is a parable Jesus tells of a man who finds a most beautiful product of biodiversity, a perfect pearl. The man then goes and sells all he has to get that pearl. Our 'all we have' can be approximated to about US$111.3 trillion per year (i.e. the gross national product of the Earth) when Costanza's 2014 study was published, and it is still less than we need for our pearl, what we need to survive and what biodiversity currently gives us for 'nothing'.

The Costanza approach and subsequent similar studies brought to public attention the new discipline of environmental economics. Current environmental economics acknowledges that, as the old song goes, 'money makes the world go round.' But it points out that our current inability to manage biodiversity sustainably has to do with the workings of economic systems that do not account for the total value and usefulness of biodiversity. That's another way of saying the 'hidden' costs are huge.

As you can imagine, Costanza and his colleagues caused a storm when those articles were published in 1997 and 2014, and the debate has rumbled down to the present day. There are issues with the calculations, the assumptions, what you include, what you exclude. But the bottom line is that, warts and all, this was the first serious attempt to value the Earth, and in a currency that those who are major consumers of biodiversity understand. However, as the authors freely acknowledge, all their calculations ignore the fact that many ecosystem services are 'literally irreplaceable'. And without such ecosystem services there would be no direct-use value.

How Biosphere 1 works – as one

In 1996 David McKay and colleagues from NASA's Johnson Space Center claimed to have found evidence of primitive bacterial life on Mars. This was in the form of microfossils and altered chemistry from a 1.9 kg meteorite, ALH84001, which landed in Antarctica 13,000 years ago. Palaeontologists were doubtful that the impressions on the rock were fossil bacteria. Similarly, there is little doubt that accounts published in early 2005 of methane production as a chemical sign of life (possibly bacterial) on the Martian surface have recently been undermined by a study that shows methane is extruded from Martian rocks. While astrobiologists continue serious attempts to look for the signature of life on asteroids and other planets, to date there is no definitive evidence of extraterrestrial life.

Earth, the Goldilocks planet – just right

Still, life does modify non-living material. And looking for chemical signs of life's activities is certainly important in any search for life, extraterrestrial or otherwise. It is a good way of investigating the impact biodiversity has on a planet with which it is intimately associated – even if that planet is just our own. Compare the atmospheric chemistry of three planets that are relatively close to one another: Venus, Earth and Mars (Table 3).

The atmospheres of Venus and Mars are nearly all carbon dioxide, with a little nitrogen and next to no oxygen or methane. If we could look at what the atmosphere of Earth was like without biodiversity, it would be remarkably similar. But add biodiversity, and oxygen now makes up about one fifth of the atmosphere being produced by plants and the like from captured sunlight. There is also an accumulation of methane from bacterial activity, and carbon dioxide is now found in trace amounts. Even though carbon dioxide is produced when many life forms respire (breathe out), it is also used up in the process of fixing the sun's energy by photosynthesis, keeping the atmospheric level low. In other words, the presence of life modifies, and is modifying, the environment of our planet in a way that has not been found on any other planet.

Table 3 Comparing the gases in the atmospheres of Venus, Earth (with and without biodiversity) and Mars

Atmospheric gas	Venus	Earth (with biodiversity)	Earth (without biodiversity)	Mars
Carbon dioxide	96.5%	0.03%	98%	95%
Nitrogen	3.5%	79%	1.9%	2.7%
Oxygen	trace amounts	21%	none	0.13%
Methane	none	trace amounts	none	none

The notion that there is an intimate, ongoing relationship between biodiversity and the physical make-up of our planet can be found in the writings of a Russian scientist from the late 1920s, Vladimir Vernadsky (1863–1945). He wrote that 'Life appears as a great, permanent and continuous infringer on the chemical "dead-hardness" of our planet's surface… [Life] is intimately related to the constitution of Earth's crust, forms part of its mechanism, and performs in this mechanism functions of paramount importance, without which it would not be able to exist.' Such a view, that to understand the workings of our planet you have to take account, and understand the role, of biodiversity, pops up time and time again in scientific writings but never really gained traction. The concept of Earth itself being, in some way, a superorganism, alive, only began to be taken seriously as science principally through the work of a British scientist.

Lovelock's Gaia hypothesis

James Lovelock was employed by NASA in the 1960s to design probes for detecting life on Mars. To do so he asked the question, 'What is it that I'm trying to detect?' He started with the fact that all living things take in energy, in the form of light, chemicals or food. They also form and get rid of waste products, and all of this they probably do using the planet's atmosphere as an exchange medium. If there were life on Mars, this should, as we discussed previously, be detectable as some sort of chemical signal. So Lovelock and another scientist, Dian Hitchcock, put together the sort of comparison of planetary atmospheres that we looked at earlier. They concluded that Mars was chemically 'dead'. Earth was not. There were still major chemical reactions going on. Lovelock suggested that 'the earth's atmosphere was an extraordinary and unstable mixture of gases, yet I knew that it was constant in composition over quite long periods of time.

Could it be that life on Earth not only made the atmosphere, but also regulated it – keeping it at a constant composition, and at a level favourable for organisms?'

Lovelock outlined his idea to his friend William Golding, of *Lord of the Flies* fame, while the pair were out walking. Golding suggested the name 'Gaia' for the idea, after the Earth goddess of Greek mythology. So in 1979 appeared Lovelock's book *Gaia: A New Look at Life on Earth*, which spelt out the view that 'the physical and chemical condition of the surface of the Earth, of the atmosphere, and of the oceans has been and is actively made fit and comfortable by the presence of life itself.'

While the idea that life made Earth 'fit and comfortable' was an attractive one, exactly *how* life did this was still far from clear. An American microbiologist, Lynn Margulis, was investigating the processes by which living things (mainly microbes living in soils) added and removed gases to and from our atmosphere. Collaborating with Margulis, Lovelock put forward the idea that there were feedback loops which regulated aspects of our planet's environment. Remember back to those science lessons in school where you may have tried to learn how a thermocouple worked? (If you never did get your head round it, this YouTube clip explains it – www.youtube.com/watch?v=qbDQ4sDLhbo.) Well, Margulis and Lovelock proposed that the same principle was operating on a planetary scale.

Let's take one topical example. Carbon dioxide is the gas produced by many living things when they respire. But it is also produced in massive quantities beneath the planet's surface and released by such structures as volcanoes. Carbon dioxide is a greenhouse gas. That is, as you increase its amount in the atmosphere, more heat is trapped and the Earth warms up. Without some way of counteracting increases in carbon dioxide, without some feedback loop, if you like, our planet would continue to heat up until life was not possible. Certainly, plants expel it at night and take it in during the day, and animals only give it out.

But by and large these exchanges balance one another, so they are not terribly effective at mopping up additional carbon dioxide. Now, one of the ways in which carbon dioxide is removed from the atmosphere is through the weathering of rocks. Water, in the form of rain, and carbon dioxide from the atmosphere react with rocks to form chemicals called carbonates (baking soda, sodium bicarbonate, is one; calcium carbonate, or chalk, is another). But vital to the regulation of carbon dioxide is the fact that this chemical process is greatly enhanced by the presence of bacteria in the soil. Importantly, the bacteria become more active as the temperature increases. So removal of carbon dioxide increases as the planet becomes hotter. The carbonates within which carbon dioxide is locked are then washed into the seas and ocean where marine life use them to make shells. Most of these shells are tiny, belonging to microscopic plants that float in the ocean as plankton. When these shelled creatures die, the shells sink to the seabed, and make up layers of what could one day become limestone. Some of the carbon dioxide trapped in these limestones will be released back into the atmosphere, as the rock can sink deep into the planet where it melts, and the carbon dioxide is released through volcanoes and so the cycle continues. Such carbon dioxide regulation is one of the many ecosystem processes, or services, we referred to earlier.

Critiques of Gaia

Criticism of Lovelock's ideas came from two main quarters. Those who thought he was making a religious statement, claiming that Earth possessed 'a life force' actively regulating all of these services. Others, who defined life in terms of natural selection (i.e. the ability to evolve), could not see how super-organism Earth could pass on its genes, so to speak. Lovelock's response to all of this was Daisyworld. Assuming that Earth was acting as a superorganism and had its own 'physiology', its own ways of maintaining itself, Lovelock and his colleague Andrew Watson proposed the Daisyworld model as a way of trying to explain how Earth could do this. On the imaginary planet of Daisyworld, there are two different types, or species, of daisy. Heat for Daisyworld comes from a nearby sun, but the planet temperature is affected and ultimately controlled by these two species. There are black daisies; they increase in number, absorbing heat from the sun and warming the planet up (demonstrating why I should not have turned up for a job interview in Nevada in my black suit!). And there are white daisies; they do the opposite, increasing in numbers if the planet starts to get too hot. Between the changing numbers of the two species, the planet's temperature remains roughly stable.

How bits of Biosphere 1 work

Gaia has been a useful hypothesis for thinking about how our planet and its inhabitants (Biosphere 1) work and interact. Noticing and predicting patterns is one thing, but actually carrying out experiments is another. And yet over the past thirty years there has been increased interest in investigating experimentally the way(s) in which biodiversity determines how ecosystems work. In particular, does manipulating the numbers and/or identities of species in a particular ecosystem affect the rates of processes or the sorts of thing that ecosystem does, contributing to the overall function of the biosphere? Such questions are interesting in their own right. How many species can we afford to lose before ecosystem function fails? Are all species equally necessary for ecosystem function, and if not, which ones are? Of course much of what is driving the non-scientific interest is the same sort of questions but with a different spin. How many, and which, species do we need to protect and conserve to keep our life-support systems functioning? How many can we afford to lose?

But first things first. Why should we think that there will be a predictable relationship between what biodiversity is and what it does? We already know that some species have a much greater influence in what ecosystems do than others. And the more species you have in your ecosystem, the more likely it is that you have a greater number of 'influential' species. And even if species were equally 'influential', they may be influential in different ways and so complement, or even positively affect, one another. So there is some reason to suspect that there is a relationship between the biodiversity of an ecosystem and how that system works. That being so, what sort of relationships would we predict?

A number of suggestions have been made about how to best describe the relationship linking the number of species in an ecosystem to the way that ecosystem works. It may be that how good the ecosystem is at doing what it is does just gets better as you add more species; so a decrease in species richness means decreased ecosystem function. This is the diversity stability hypothesis (Fig. 24C). The second suggestion is that a loss of species will not affect how the ecosystem works, at least down to a critical level. Those remaining can make up for the losses, although there is a minimum number of species needed to make the whole thing work. This is referred to as the redundancy hypothesis (Fig. 24A). The third suggestion is that losing a few species may seem to have little effect, but the more you lose, the greater the detrimental effect on ecosystem function. This is termed the rivet-popping hypothesis (Fig. 24B), with the analogy being that species in an ecosystem are like rivets holding an aeroplane together. The fourth suggestion is that a small number of particular species are key to an ecosystem. In this case loss of just one species could result in a dramatic loss in ecosystem function. This is the keystone species hypothesis (Fig. 24D). The fifth and final suggestion is that the world is a complicated place, and species have numerous complicated roles and interactions, so that while a change in biodiversity will result in changes in function, you don't have a snowball's chance in Hades of putting forward one overall theory that will explain everything. Here we have the aptly named idiosyncratic hypothesis (Fig. 24★). That's not to say the relationship is necessarily unpredictable. For instance, it may be that the removal of one key species will have a different effect from removing other species in the ecosystem. A species with such a pivotal role in how an ecosystem is put together is referred to as a keystone species.

And finally, despite what we said earlier about having good reason to expect a relationship, there is still the sixth possibility – that there is just no relationship.

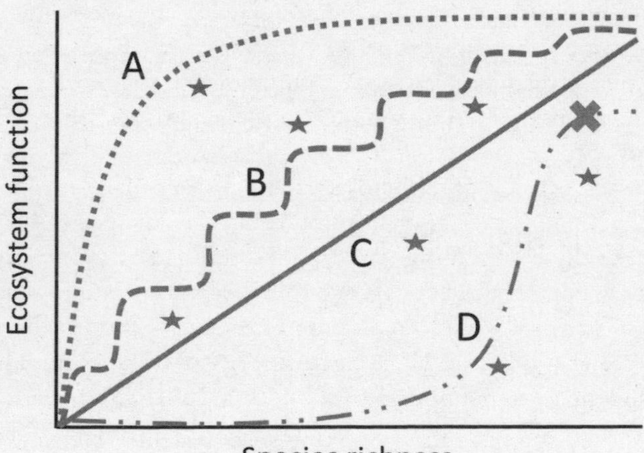

Figure 24 Early ideas on possible relationships between biodiversity (species richness) and ecosystem function (A. redundancy hypothesis; B. rivet-popping hypothesis; C. diversity–stability hypothesis; D. keystone species hypothesis; stars – idiosyncratic hypothesis.)

Where have we got to in testing these ideas, these hypotheses? Much of the initial experimental work involved looking at plant (often grass) species richness and relating that to productivity. Dave Tilman and his team, for example, took a large number of fields where he manipulated the numbers of grass species present. BIODEPTH was a European project, with eight very different sites, that set out either to confirm or refute the existence of a relationship between species richness and productivity in grasslands. In each case, and with all the studies that have been carried out to date, the results are complicated. Interpreting the results has been made difficult by the way in which some of these very complicated experiments have been designed, and there is considerable debate in the scientific literature as to what can and cannot be deduced from them.

The period of investigating biodiversity and ecosystem function in the way we've described continues, but over the past ten years new approaches have emerged. More emphasis has been put on incorporating multifunctionality – that is, not just looking at ecosystem functions in isolation – and trying to think about dynamic communities rather than the static ones which are certainly easier to work with in controlled experiments. In some ways our understanding has advanced. It does seem overall that if you increase the total number of species present, you increase the number of redundant species, even though exceptions have been identified. But this is far from the whole story. It is complicated by the fact that changing the conditions within an ecosystem may change the identity of the 'influential' species present. So it is not just how many species make up an ecosystem that is the issue but instead making sure that the ecosystem can draw on a pool of species (and often the right species in terms of what they do) to act as a sort of buffer to change. You will have gathered by now that we are still a long way from identifying some grand unifying theory. But we have made a good start. Taking a broad-brush approach, all of the experiments carried out to date have tended to confirm what biologists have thought for years, and rural communities have known for centuries, that greater biodiversity, on the whole, is better for the functioning and stability of ecosystems.

Another complementary approach to understanding how ecosystems work is to attempt to build your own. This is what we explore next.

Build your own biosphere: Not-so-silent running

The home marine aquarium

Several years ago I had a few hours to kill at Charlotte Airport in the US. In one of the airport shops I came across a small, hand-blown glass ball for sale. Completely sealed off from the outside world, it contained a couple of tiny red shrimps, a beautiful piece of gorgonian coral and some microbes, all living in seawater. This was retailing as an executive toy, the Ecosphere. All it needed from outside was a certain amount of sunlight. A spin-off from NASA research into sustainable ecosystems in outer space, this was a self-sustaining system that would last for about three years. The associated blurb claimed that some could last as long as ten. An isolated, almost self-sufficient world in the palm of your hand.

One of the first serious attempts to construct a self-supporting ecosystem was born out of the nineteenth-century idea of having a low-maintenance aquarium in English middle-class homes. A certain 'gentleman', Mr Robert Warrington, presented some early experiments to the British Chemical Society in 1850. His scientific paper was entitled 'On the adjustments of the relations between animal and vegetable Kingdoms, by which the vital functions of both are permanently maintained'.

He told the assembled gathering that he placed two goldfish in twenty gallons of spring water, which half-filled a glass bowl. Some sand, mud, pebbles and bits of limestone were carefully added to carpet the bottom of the bowl. The mud was used to hold the roots of a small water plant called *Vallisneria*. Warrington then left the aquarium undisturbed until a green scum started to coat the walls of the bowl, obscuring his view of the goldfish, and the goldfishes' view of him. The water began to cloud over and look distinctly unhealthy. The addition of some snails, which fed both on the green scum and some of the decaying vegetable matter, resulted in the water clearing in a very short space of time. Minor pruning by Warrington kept the plants happy and healthy, the snails flourished on shed leaves, and the fish grew fat and healthy on snail eggs for many months.

A mere three years later he had published a number of essays on how to set up a self-sustaining marine aquarium. This was probably driven by the beginning of a Victorian love affair with the seaside. Numerous natural historians, including Philip Henry Gosse of *Father and Son* fame, got involved in producing what were, in effect, partially self-sustaining ecosystems. Gosse was drawn into the construction of London's first large-scale public aquarium, in the Gardens of the Zoological Society. Through books such as his *Aquarium* and *Devonshire Rambles*, he sparked off a craze in keeping marine aquaria in private houses, often many miles from the sea. This was the first commercial attempt to sell the idea of self-sustaining ecosystems (or closed ecological systems, as they have come to be known).

Using closed ecological systems as a research tool has largely been overlooked since those days of the first marine aquaria. The beginning of serious research can be traced to the early 1960s and has evolved from using small glass bottles right through to the Biosphere 2 project and the NASA-funded Closed Ecological Life Support System (CELSS) project.

Mysteries and hazards

Silent Running (1972) is a cult science-fiction movie. It presents Freeman Lowell, a nature-loving astronaut, aboard the *Valley Forge*, a gigantic spaceship which is part of a small fleet. Their mission is to save the last surviving forests of an overpopulated and environmentally degraded Earth. The film ends with three little pre-R2-D2 robot characters tending, watering and generally looking after the forests as they floated endlessly though space. Here was a self-sustaining ecosystem, a little biosphere that would have been capable of sustaining Lowell and the other three crew, had Lowell not disposed of his companions early on in the film when they wanted to dump the forests and return to Earth.

Keeping a bunch of marine animals and plants healthy for a couple of months or even years is one thing, but building an ecosystem that can support human life? How close are we to science fiction becoming science fact?

Biosphere 2 (remember, Earth is Biosphere 1) was an airtight facility of 204,000 cubic metres built between 1987 and 1989. Seven years in construction, it was completed in 1991 at an estimated cost of US$200 million. Lots of small bits of large natural systems – coral reefs and grasslands, for example – were crammed together in what looked, from a distance, like a huge greenhouse. The 'experiment' consisted of enclosing eight people inside and seeing if the human-made biological system could supply their food, water and air needs for two years (1991–93). It is difficult to know how many of the 'challenges' to the lives of the biospherians, as they were called, could have been predicted. The microbes in the soil that was used to grow food consumed much more oxygen and produced much more carbon dioxide than was expected. A good proportion of the carbon dioxide reacted chemically with calcium-containing concrete used in the construction of the biosphere. The chemical result of this reaction was chalk. The atmospheric result was that a lot of the oxygen contained in carbon dioxide that was needed to be transformed back into breathable oxygen by the plants was, in fact, locked tight in the structure of the building. So during the 'lock-in', breathable oxygen had fallen to two thirds of its normal level (equivalent to living at the top of a very high mountain) and carbon dioxide was high and all over the place. Just in case you're holding your breath by this time, don't worry, well before the two years were up the outside 'backup' team added oxygen to the biosphere atmosphere from outside. But even this did not mean that the biospherians could breathe easily. Even the plants introduced to recycle 'waste' carbon dioxide had their own surprises. Some of the vines turned out to be very competitive. Vigorous weed management failed to stop species like morning glory from threatening huge areas laid aside for food production. Bits, and often big bits, fell off big trees, and nineteen of the twenty-five (non-human) vertebrates became extinct. So too did many of the insects and all the pollinators. Plants that rely on pollinators for reproduction were technically extinct – they just hadn't caught on to the fact. Wonderfully named crazy ants did really well, as did a few cockroaches. Aquatic environments became polluted because of nutrient overload – just like Biosphere 1.

The bottom line was that the eight people within Biosphere 2 were quite literally involved in a life or death struggle every day. A struggle to maintain services from nature that we enjoy daily, take for granted, and to a large extent, get for free. Despite almost unlimited backup from the outside, it was impossible to maintain a closed system that supplied the material needs of just eight individuals for two years. In the words of Joel Cohen and Dave Tilman, 'At present there is no demonstrated alternative to maintaining the viability of Earth. No one yet knows how to engineer systems that provide humans with the life-supporting services that natural ecosystems produce for free... Despite its mysteries and hazards, planet Earth remains the only known home that can sustain life.'

Valuable for what, and to whom?

The whole notion of non-use value was first put forward in 1967 by John V Krutilla of Resources for the Future. Although differing a little bit, depending on what books you read, there are essentially three non-use values: option, bequest, and existence (Fig. 23, p. 167).

Keeping options open

Option (or potential) value has been described as the value of something that has not yet been recognised or, as Joni Mitchell put it more poetically, 'You don't know what you've got 'til it's gone/You pave paradise and put up a parking lot.' Once biodiversity is lost the option of benefiting from it, even if you don't currently know what these benefits might be, is gone for ever. You could think of option value as like an insurance premium that you are willing to pay to retain the option of future use on goods or services just in case. One of the best illustrations of option value is the seed bank collection of the International Rice Research Institute set up in 1961.

We've already mentioned that rice is one of the most important food plants on the planet. It provides sustenance for more than half the world's population. In the 1970s rice crops were infected by the grassy stunt virus, transmitted by the brown plant hopper that at the time infested huge rice-growing areas in Asia. The high-yield varieties of rice in use had no resistance to the virus. The International Rice Research Institute maintained a seed bank representing many thousands of rice varieties, including recently developed hybrids and strains, as well as wild varieties that were no longer used in cultivation. They tested in excess of six thousand different rices for resistance to the virus. Only one, a strain no longer in use, was resistant. In 1976 this redundant strain was crossbred with others and a new strain created, known as IR36. IR36 is both a high-yield rice and is resistant to the virus.

The take-home message from this story is that without access to a rice strain thought to be redundant, the economies and lives of many, many people would have been quite different, and possibly negatively so. In this case, and in that of many hundreds of others, genetic information was viewed as a source of novelty, with extinction literally being an irrevocable loss of that information.

Bequest and bequeathal

Scottish philosopher-economist John Locke was one of the first to formulate the idea that each generation should bequeath 'enough and as good…for others' to future generations as justice demands it. This is termed bequest value. A common example is that of preserving a national park. Even though not all of the present generation will have the intention of visiting the park, let alone use and enjoy it, its preservation will benefit future generations. But bequest value is much more than just leaving something behind for future generations. It encompasses the notion of compensating future generations for the loss of biodiversity value – be it money, production, ecosystem services – we are responsible for right now. In other words, it is recompensing our children and their children for our use of what should have been their biodiversity. For example, economists sometimes use surveys to estimate the monetary value of bequest value by asking questions such as 'How much money would you be willing to pay to save, say, the tiger or the panda, or a piece of land for future generations?'

There is one more set of non-use values we need to consider: existence and intrinsic. These are notoriously difficult to disentangle and describe. Existence value could be defined as the value we, as people, place on just knowing a species, a habitat or an ecosystem exists, even if we will never see it or use it. To convert this intangible value into hard cash, we could go back to our economist with their survey, and ask people how much money they would be willing to pay just for the knowledge that the tiger, the panda or the piece of wilderness exists (existence value). This is one, but only one, way of doing it. Part of the problem with the approach, whether we're talking about bequest or existence value, is that such questions are always hypothetical (Q: 'If you had to, how would you murder your partner?' A: 'Well, I would…').

For the remainder of this chapter we'll trace the history of some of the ideas, philosophical and religious, that have informed what we think about intrinsic and existence value and see how those ideas have shaped our present-day world views of biodiversity.

Full-on philosophers and laid-back religion?

Value bestowed, not intrinsic

Throughout our history, biodiversity, and nature generally, has furnished us with the raw materials and the inspiration for our existence and our lives. It should not be surprising, therefore, that much of the value of biodiversity has been seen to be tied up with our own existence, well-being, cultural activities and spiritual lives. It was perhaps summed up most succinctly by John Locke, whom we met earlier. He claimed that everything in nature was waste until we transformed it into something of value. There is not necessarily any intrinsic value – something is only valuable to the extent that it satisfies our physical, mental and spiritual needs. Arguably, this has been the most influential idea and assumption, dominating Western thought from classical times.

The Greek tragedian Sophocles (496–406 BCE), the polymath Aristotle (384–322 BCE) and Roman stoic Epictetus (55–135 CE) all propounded ideas on the excellence of human nature, and the precedence of humans over all other species. Sophocles even went as far as to say that humankind 'is master of ageless earth, to his own will bending… He is Lord of living things.' This belief in human superiority and domination of nature was a common theme in medieval and Renaissance thought despite a few dissenters. The Church taught that God's commands in the book of Genesis, to dominate and subdue, meant that nature was there for us to do as we will. We were its masters. This is clearly seen in Thomas Aquinas's *Summa Contra Gentiles*, where he says, 'Hereby is refuted the error of those who say it is sinful for a man to kill dumb animals: for by divine providence they are intended for man's use in the natural order. Hence it is no wrong for man to make use of them, either by killing or in any other way whatever.'

Human superiority and domination were no less prevalent in secular thinking, and to some extent were made easier by the views emerging from Renaissance scientists and thinkers, like Descartes and Galileo, of nature as a machine. Conquest and subjugation of nature in different forms continued to dominate thoughts on our relationship with biodiversity. This has continued to the present day, even if ideas such as Darwin's theory of natural selection, and evolution generally, have done much to show our own 'less exalted' place in the natural order. Humans had become more thoroughly part of biodiversity but this did not stop them lording over it. Some philosophers like Immanuel Kant (1724–1804) in some ways embodied the prevalent view that nature had no purpose, so he suggested that we act as if it had. In the twentieth century, philosopher Jean-Paul Sartre (1905–80), biologist Jacques Lucien Monod (1910–76) and philosopher Bertrand Russell (1872–1970) all held the view that nature had no value except that which we project on it. Perhaps the most extreme view was that of Friedrich Nietzsche (1844–1900), who tried to destroy the ideas of nature as a machine, as a living being, as having any purpose or any beauty and harmony: 'The total nature of the world is...to all eternity chaos... The living being is only a species of the dead, and a very rare species.'

Much of human history has seen the purpose of economic growth and technological advancement as to maintain and enhance our material well-being. Thus, the whole idea of conservation in this world view is one of efficient management of limited resources, converting wild and hostile nature into a more benign, more 'comfortable' environment.

Intrinsic value

Petrarch (1304–74) believed that nature was a sign of God's providence, but he also thought that it existed for its own sake. In his book *The Ascent of Mont Ventoux*, he writes, 'Today I ascended the highest mountain of this region...I admired every detail.' German explorer and biologist Alexander von Humboldt (1769–1859) declared his excitement and ascetic delight when encountering the natural world. Aldo Leopold (1887–1948), an American naturalist and the 'father' of the field of wildlife ecology, was one of the most famous supporters of the idea that wildlife and wildlands were valuable in and of themselves. He wrote: 'The last word in ignorance is the man who says of an animal or a plant: "What good is it?" If the land mechanism as a whole is good, then every part is good, whether we understand it or not. If the biota, in the course of aeons, has built something we like but do not understand, then who but a fool would discard seemingly useless parts? To keep every cog and wheel is the first precaution of intelligent tinkering.'

Those who would question the existence of any notion of intrinsic value point out that Petrarch, and all the others who say they enjoy nature for its own sake, still have the major emphasis on their enjoyment and not the intrinsic value. Despite this type of objection, we still have reference to the intrinsic worth of biodiversity contained in many of the treaties and conventions drawn up over the past decades, using such value as one of the bases for conservation. Deep ecology is a world view that has emerged over the past few years, which has at its centre the recognition of nature's right to exist quite apart from any benefit we may derive from it. So, despite the fact that it is difficult to demonstrate, the view that biodiversity has intrinsic value seems to have grown over the last hundred years or so. However, even those who hold that biodiversity and nature have their own right to exist, and should be conserved for their own sake, acknowledge that our own future and that of biodiversity are inextricably intertwined.

Valued as an object of worship or through kinship

The worship of nature, which includes biodiversity, is referred to as pantheism. Pantheism is a persistent and pervasive feature of human history and belief. In the oldest Hindu scriptures, the Vedas, which are 3,500 years old, nature is worshipped and revered. The influential Portuguese-Jewish philosopher Spinoza (1632–77) believed in the divinisation of nature. Today this is found in many forms, from old established religions, such as paganism, through to 'new age' splinter groups. There is a version of the Gaia hypothesis where Gaia as a biological theory imperceptibly morphs into the Earth mother of mythology. There are even 'Christian' forms, such as the process theology of philosopher AN Whitehead (1861–1947), where God is evolving as part of his creation.

Perhaps more common than the divinisation of nature is the world view that humans are very much an inextricable and *equal* part of biodiversity. This is undeniably true in a biological sense – all life is related – but the idea that animals in particular are our kin takes this a stage further, claiming an even greater unity. Taoism, as embodied in the writings of the mystic Lao-Tze (sixth century BCE), emphasised the essential unity of humanity and nature. The same sort of belief was held by many indigenous peoples, such as Native Americans. The Lakota nation believed that 'all things were kindred and brought together by the great mystery.' Schopenhauer (1788–1860) was one of the first modern philosophers to attack the notion that humankind is better than nature.

The Scottish-American naturalist John Muir (1838–1914) believed in a harmony between humankind and nature, spiritual as well as physical. This he shared with other American writers and thinkers of the nineteenth century, such Ralph Waldo Emerson and Henry Thoreau. Muir loved wilderness and was a huge influence in promoting the idea of forestry conservation and setting up national parks. In many ways he is seen as the first person to develop a modern environmental ethic. In *The Wild Parks and Forest Reservations of the West*, he wrote, 'Thousands of tired, nerve-shaken, over-civilized people are beginning to find out that going to the mountains is going home; that wildness is a necessity; and that mountain parks and reservations are useful not only as fountains of timber and irrigating rivers, but as fountains of life.' The modern version of this unity or kinship is what is known as the biophilia hypothesis.

Harvard biologist Edward O Wilson is regarded by many as the father of modern biodiversity and has perhaps done more than anyone else to bring the importance of biodiversity to popular attention. He published a slightly different type of book in 1984 entitled *Biophilia: The human bond with other species*. Here he suggested that human beings, as part of the natural world for many hundreds of thousands of years, have a natural, innate regard and need for living things. This Wilson called biophilia, which he defined as 'the connections that human beings subconsciously seek with the rest of life'. Other advocates of this emotional or spiritual dimension to our relationship with biodiversity have highlighted studies that show patients recovered quicker if they were exposed to greenery, even images of greenery, compared to recovery in an artificial environment. Some have even linked biophilia with the Gaia hypothesis (humans are one element of Gaia) and not infrequently such discussions border on the divinisation of nature that we started with.

A creator gives biodiversity value

It is true that for much of our history there has been a religious as well as secular emphasis on the superiority of humankind and the value of nature solely in terms of satisfying human needs and desires. That said, the reasons for this are not always as straightforward as they are made out to be by many modern commentators. Much of the blame for the over-exploitation of biodiversity, and nature in general, is laid at the door of religious belief. In particular, those faiths that seem to draw a firm distinction between humans and the rest of biodiversity – Judaism, Christianity and Islam – come in for criticism as being biodiversity-unfriendly, while other more biodiversity-friendly belief systems like Hinduism, Buddhism and Taoism are complimented. However, there are two points that should be made.

First, whether your country or region or culture is characterised by a biodiversity-friendly or allegedly biodiversity-unfriendly belief system is not clearly reflected in the ways that regions or cultures, in the past or today, treat biodiversity. The widely held view that peoples whose religions require them to live in close contact with nature and to respect it have not been responsible for major historical extinctions has very little factual basis. The bottom line is that most human beings, irrespective of colour or creed, throughout their history have treated biodiversity – on different scales admittedly – rather badly, whether they were aware of it or not.

Second, just because Judaism, Christianity and Islam hold that there is a qualitative difference between humans and the rest of biodiversity, it does not necessarily follow that they should treat biodiversity badly. That may happen, but not as a result of the beliefs of those faiths. In fact, the very opposite should be true. The whole tone of the creation account in the book of Genesis (and many other sections of what Christians have traditionally referred to as the Old Testament), where all three faiths to different extents start with their 'environmental ethic', is one of stewardship of Earth and its resources. Humans are not the measure of all things biodiversity. While they are considered different (and special) from biodiversity as a result of individually bearing the image of God, these writings also make it clear that we were created from the same stuff (dust, earth) and eat and reproduce as the animals do. This is what led Blaise Pascal (1623–62) to emphasise the humble place of humankind at a time (the Renaissance) when both religious and secular thought seemed bent upon the 'divinisation of man'. Humans are *not* the master of biodiversity in these sacred writings. They are portrayed as being charged with looking after the planet on behalf of someone else – and that someone else is the creator of the planet and all that was on it. And he declared it was good. We are depicted as stewards – not owners – and God is the creator, sustainer and owner of biodiversity. In the management process, humankind were granted leave to use biodiversity for their own subsistence, but this is a far cry from Aquinas's 'do whatever you want with biodiversity.' The English translations of the book of Genesis, which talk about 'having dominion over' and 'subduing' the Earth, convey a very different sense from the stewardship theme that runs through the scriptures of these three world faiths. In the Judaeo-Christian scriptures there are laws about the treatment of land

and biodiversity. In one of them, soldiers are forbidden to cut down fruit trees in war, even if their lives depend on it. Another theme running through all three faiths is the view that biodiversity and the natural world glorifies and reveals God. Although this notion has a long religious history, it was actually brought to bear on scientific study, as its motive and impetus, from the Renaissance up until Darwin's time. John Ray (1627–1705) and William Paley (1743–1805) both thought nature worthy of study and attention because it revealed God.

One of the most influential Christian theologians of the twentieth century, Karl Barth (1886–1968), believed that we had a moral responsibility for nature. Barth was very wary of any notion of 'reverence for life', but believed that 'the world of animals and plants forms the indispensable living background to the living-space divinely allotted to humans and placed under their control. As they live, so can humankind. Humans are not set up as lords over the Earth, but as lords on the Earth, which is already furnished with these creatures. Animals and plants do not belong to humans; they and the whole Earth can belong only to God. But he takes precedence of them.'

This Barthian view is to some extent echoed and expanded in Pope Francis's response to our biodiversity and environmental crisis, *Laudato Si'. On Care for Our Common Home* (2015), and by a number of Eastern Orthodox, evangelical and Protestant groups (e.g. the issuance and endorsement of *An Evangelical Declaration on the Care of Creation*, and the extension of the Ethiopian Orthodox Church forests to protect the country's fragile landscape and help reverse massive deforestation). They stand in stark contrast, at least in practice, to the anti-environmentalism that seems to be issuing from many vocal, right-wing American evangelists at this time. But most Christian denominations seem to have responded well to Francis's message, although admittedly there have been critics both within and without the Catholic Church. Francis concludes that 'a true ecological approach always becomes a social approach;

it must integrate questions of justice in debates on the environment, so as to hear both the cry of the Earth and the cry of the poor… Humanity still has the ability to work together in building our common home.'

To conclude

What should be very clear from this chapter is that biodiversity has value, and value over and above its direct use values. Biodiversity is at least worth more than our current global economy. This value may be difficult to categorise and pin down but few would deny its *worth*. The prevalent view is that, either from a religious or a secular point of view, with different motivations, we have a moral responsibility to protect the life forms with which we share this planet. With this in mind we move on now to look at the steps we, as a global community, have taken to protect and maintain this biodiversity that we, in theory at least, value so much.

7

Our greatest hazard and our only hope?

> Fearful and unprepared, we have assumed lordship over the life or death of the whole world – of all living things. [We have] become our greatest hazard and our only hope.
>
> John Steinbeck, Nobel Banquet Hall, 1962

Saving private land

Bird Rock beach is situated within the South La Jolla State Marine Reserve (designated in 2012) and just south of the San Diego–La Jolla Marine Conservation Area (originally Ecological Reserve), a protected area consisting of 2.16 square kilometres of waters just off La Jolla Shores and La Jolla Cove. Although small compared with other marine reserves worldwide, the latter, designated in 1971, is one of the oldest 'no-take zones' in Southern California. It was established in an area that previously had been heavily impacted by coastal development and overfishing. The goal set for the reserve was 'to protect threatened or endangered native plants, wildlife, or aquatic organisms or specialised habitat types'. Thirty years later a study evaluating the extent to which this goal had been achieved was published by researchers from Scripps Institution of Oceanography. They concluded that while there were individual success stories (e.g. enhancement of marine resources), the reserve was probably too small to be effective for most harvested species. Evaluating the extent to which the conservation area goals had been achieved was hampered by a lack of information on what lived in the area before 1972, for comparison with what lives there now. However, a number of new marine conservation areas have been established both in the La Jolla district and Southern California more generally.

A conservation partial-success story. And there are hundreds if not thousands of such stories worldwide. But it would be folly to believe that Bird Rock and its immediate surroundings are now safe. What happens in San Diego, in California, in the Pacific Ocean to the west of the beach, in the US, in the northern hemisphere, globally, all affects efforts to conserve the biodiversity of Bird Rock. Without action at an international and global scale, our very best efforts to protect the biodiversity of Bird Rock could turn out to be as significant as rearranging deckchairs on the *Titanic*.

To a large extent the academic subject of biodiversity was born out of the need to gather information for conservation purposes. So, as we approach the end of this book, we will attempt to trace the main global conservation issues of the past and assess their state now, in the present (Fig. 25). As mentioned earlier, we could pack pages and pages with stories of local biodiversity conservation success stories (and probably an equal number of failures). But given the global nature of biodiversity, this international level is what we will deal with here. That is not to say that the local stuff isn't important. Without it, perhaps, we would have no biodiversity to protect globally. 'Either it all is important or none of it is.'

Antecedents

Although the origins of the modern conservation movement can be traced to the mid nineteenth century, it was only after the

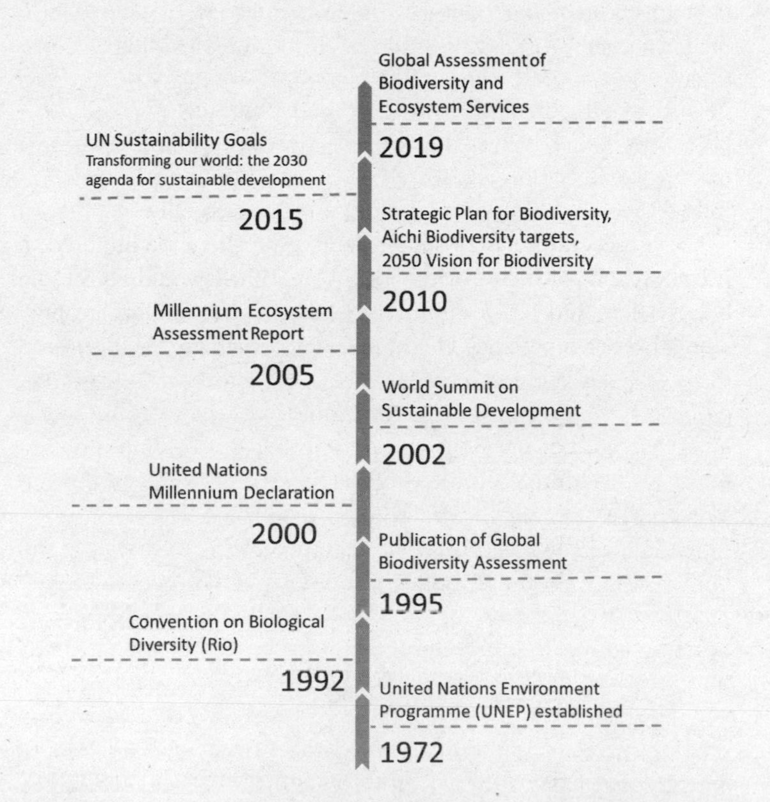

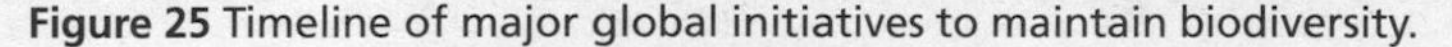

Figure 25 Timeline of major global initiatives to maintain biodiversity.

Second World War in the twentieth century that we began to pick up global concerns about biodiversity. In 1972 the United Nations convened the first global conference ever to address environmental issues. That meeting stimulated the formation of environmental ministries and civil organisations promoting environmental issues across the world. The United Nations Environment Programme (UNEP) was established. There was the explicit acknowledgement that the forces driving the destruction of the natural world lay outside of the remit of biology, and so biology would not be enough to achieve some degree of protection. Integrating science into the bigger picture that included politics, economics, morality, social responsibility and so on was seen as key to conservation at every level.

The word biodiversity, an abbreviated form of biological diversity, was coined by Walter Rosen, the co-director of a 1986 conference entitled 'The National Forum for Biodiversity'. The book that resulted from this conference was *Biodiversity* (1988), edited by EO Wilson. To a large extent this book sparked off both the concept and the use of the word 'biodiversity' in a way that continues even today. Wilson's beautifully written, more popular book *The Diversity of Life* (1992) did much to bring biodiversity into popular thought and consciousness. Pressure from UNEP and a large number of environmental groups, together with a growing awareness of biodiversity and its loss in the mind of the public, resulted in the first global attempt to conserve biodiversity in Rio de Janeiro at the beginning of the 1990s.

Oh, Rio

The Convention on Biological Diversity (CBD) was a landmark document in our attempts to look after biodiversity. Signed by more than 150 nations on 5 June 1992, the treaty arose from the United Nations Conference on Environment and Development, held in Rio de Janeiro, Brazil. It came into force about 18 months later. On the plus side, and not to be underestimated, this was the first attempt at a truly global treaty, and it emphasised more than ever before that the conservation of biodiversity was a common concern, no matter who you were or where you lived. Genetic diversity was included explicitly for the first time as a conservation priority. On the downside, many countries were slow to ratify the document and even now a number of the key players have yet to fully come to the party.

Despite its complicated legal jargon and multiple caveats, the ideas that are core to the treaty are relatively easy to grasp. The objectives of the CBD are: 1) the conservation of biological diversity, 2) sustainable use of its components, and 3) equitable sharing of the benefits arising from the utilisation of genetic resources. The bottom line is that conservation, sustainable use, and sharing of benefits must take place, if only because otherwise we imperil our own existence. There are 42 articles in the CBD. They range from carefully worded aims, objectives and definitions, through to policies that need to be implemented, but they also spell out what the signatories have actually signed up to. Overall, it's a lot more than setting up a few nature reserves and bringing a couple of charismatic creatures back from the edge of extinction. In what follows we go through the main points of the CBD and also, to some extent, use this as a framework for introducing basic conservation concepts such as *in situ* and *ex situ* conservation.

Large brushstrokes

The first article gives the broad-brush picture of what the whole convention is about – conserving biodiversity, using its constituent parts in a sustainable way, and doing so in the context of 'fair and equitable sharing'. There's biology here, yes, but set firmly in a social, political and global context. All very laudable. But how is this to be achieved? Conserving biodiversity is not going to happen purely because a hundred or so countries sign up to a treaty that says that biodiversity is a 'good thing'. As we have seen, there is little encouragement from our track record on maintaining biodiversity. As the British psychologist Havelock Ellis said, 'The sun, the moon and the stars would have disappeared long ago had they happened to be within the reach of predatory human hands.' So signature nations of the CBD were obliged to put into place strategies, tangible ways by which biodiversity could be conserved and used in a sustainable fashion (Article 6).

Louder than words

Article 6 is arguably the most far-reaching and significant article. For it to be effective, sustainability must touch on and include almost every area of a nation's activities. Let us be clear about this. Even if only Article 6 on sustainability were taken semi-seriously, it would fundamentally change the way nations are governed. One of the more obvious things that would need to change is perverse subsidies – financial support given to processes that in the long run have an adverse effect both on biodiversity and the economy. Support for fossil fuels, which increases pollution, negatively affects human and animal health, and contributes to global warming, is a good example of a perverse subsidy. Some economists have estimated that the amount of money financing perverse subsidies often exceeds the marketable value of the goods being generated.

Article 6 asks that strategies and plans be drawn up that will affect biodiversity conservation on the ground. A tangible way we can see how this works is the article's effect on the management of Dartmoor National Park, quite close to where I currently live in Plymouth (no, not Massachusetts, the other one, in Devon, England). As a result of the Rio summit in 1992, all the signatory nations had to put national action plans into place – documents that identified, at the national level, biodiversity priorities. The UK Biodiversity Action Plan was launched towards the end of 1994, followed shortly afterwards in 1995 by the setting up of the UK Biodiversity Steering Group. There followed the production of a series of habitat and action plans for different parts of the country. *Action for Biodiversity in the South West*, published at the beginning of 1997, started the planning process for a number of counties, including Devon. About a year later, *The Nature of Devon: A Biodiversity Action Plan* took the planning at local level a stage further, while English Nature and Dartmoor National Park Authority produced *The Nature of Dartmoor: A Biodiversity Profile*. By early 2001 the *Dartmoor Biodiversity Action Plan* had been produced and in September the Dartmoor Biodiversity Project was launched. Similar implementations of the CBD were going on all over the world, admittedly with different degrees of success and/or haste, but still they were happening. Depressingly, many are still mere aspirations. Regularly, there appears to be a failure to recognise the fundamental nature of exactly what needs to be done.

Putting a Biodiversity Action Plan or similar in place is all fine and good. But how and when do we know that we are close to achieving what we set out to do? The CBD recommends a carefully thought out programme of monitoring (although it did not say how this was to be done). In effect, an information gathering exercise, an audit of biodiversity and of the mechanisms put in place to conserve it and use it sustainably. We've already seen how difficult it is to get an idea of *just* the number of species on Earth, never mind anything more complicated. This degree of realism is shared by the CBD, which suggests that such information gathering be targeted on components of biodiversity believed to be key for conservation and sustainable use, and threats to those components.

Arks in parks

One of the longer and tortuously complicated articles is Article 8 on *in situ* conservation. Conserving things *where they are*. This obliges signatories to set aside a number of protected areas within their national boundaries, protect the land that borders protected areas, attempt to restore or aid the recovery of degraded habitats, and combat the risks associated with alien species. We'll say more about the establishment of protected areas later, but in the meantime we should note, in passing, that difficulties in providing financial and other support for *in situ* conservation, particularly to lower-income countries, is a recurrent theme of conservation. Later sections urge signatories to look for ways of minimising conflicts between conservation and present use, while the final sections highlight the legal and financial obligations of such conservation, recognising that lower-income countries will need help.

Out of place – but alive

As well as conserving things where they are (*in situ*) allowance is also made in Article 9 of the CBD for conserving species outside of their natural habitats (*ex situ* conservation). While *ex situ* is always seen as a second best to *in situ*, even second best is as good as it gets sometimes. As we saw earlier, currently there are thousands of botanic gardens and zoological parks (depending on how you define them) worldwide. They could be the key to *ex situ* conservation in terms of breeding, holding and reintroducing endangered species. There are also more specialised biodiversity 'holding facilities' such as sperm/ova banks, culture tissues and seed banks (like the International Rice Research Institute, which we came across in the previous chapter). The case for zoos as effective places for the conservation of biodiversity is still hotly debated. It has been pointed out that of the million or so individual vertebrates held in zoos very few of them have any great conservation status. In the case of mammals, it's been estimated at about one in ten. However, supporters of zoos point out that there are over three hundred endangered species in captivity, and they have successfully preserved species such as the Panamanian golden frog and Przewalski's horse from total extinction. Certainly, the earliest *ex situ* success story must be Père David's deer, which became extinct in the wild in China three thousand years before the present but survived in an area of parkland. *Ex situ* is often seen as complementary to *in situ* conservation as it is sometimes possible to take species from captivity and reintroduce them back into the wild.

Buzzword for the twenty-first century

'Sustainability' must rank as one of the key buzzwords of the twenty-first century. The sustainable use of the components of biodiversity is what Article 10 is all about. Given that, as we've seen throughout, most uses of biodiversity have not been (and are not) sustainable and that currently human exploitation is estimated to be 20% greater than the Earth's productivity, this notion has an air of urgency about it.

Sustainable development is all about reconciling economic development with environmental protection. It's meeting the needs of the present with an eye on protecting our natural heritage for our children and their children. For example, in the second half of the twentieth century, cultivation of high-yield crops, with the intensive use of fertiliser and pesticides, resulted in considerable increases in productivity. More crops for each buck you spent. But putting values to many of the hidden costs suggests that this was not as good a deal as we thought we were getting. The environmental and biodiversity costs were substantial. What was needed was a compromise between what was truly economically viable, and technically possible, and what was biologically acceptable. Today, taking sustainability issues more into account, cultivation systems are becoming more diversified, as is what farmers do and how they do it.

For our activities to be truly sustainable there has to be a major cultural shift. And this requires the support of local peoples.

It is not a case of going back to some 'golden age' when we allegedly 'lived in harmony' (sustainably) with nature. As already alluded to in previous chapters, primitive humankind had a pretty major negative impact on biodiversity, and even where there were examples of 'living in harmony', they only worked because of low human density and the absence of commercial exploitation. There is no golden age to go back to, and even if there was, we couldn't go back now anyway.

Article 11 is all about incentive measures. Signatories should adopt economically and socially sound measures that act as incentives for conservation/sustainable use. As biodiversity loss is driven by economics, we should harness the same forces to protect biodiversity. However, in reality very often the converse is true, as the abundance of perverse subsidies show. The remaining articles deal largely with procedural matters, and while obviously crucial, reading them makes watching paint dry a pleasant alternative.

Responses to Rio

One of the potentially major outcomes of Rio was Agenda 21. Basically, in 1992 the international community signed up for what was billed as an unprecedented global plan of action for sustainable development. This meant both improving people's lives and conserving natural resources against the backdrop of huge population growth and the accompanying increased demands for economic security, energy, food, healthcare, shelter, sanitation and water. To this end, the UN Commission for Sustainable Development was set up in 1992. Its job was to monitor and provide feedback on how Agenda 21 was being implemented on the ground at local, regional and international levels. The first major, independent, peer-reviewed analytical review of the biological and sociological aspects of biodiversity, *The Global Biodiversity Assessment*, was published by UNEP in 1995.

However, by 2002 it was widely acknowledged that progress on implementing Agenda 21, and the other action points from Rio, had been extremely disappointing, with poverty worsening and destruction of the environment carrying on, and in some areas even accelerating, at alarming rates. In the summer of 2002 a list of the top ten sustainable development successes and failures since Rio, were published by the International Institute for Sustainable Development and for many it made disappointing reading.

It was recognised by the UN that what was wanted (and needed) was not philosophical or political talking shops but rather a set of concrete plans and actions, and ways of making those plans happen. In response to this need, the Commission for Sustainable Development organised a follow-up to Rio. This was the World Summit on Sustainable Development, held just outside the city of Johannesburg, 26 August–4 September 2002. With over 23,000 people participating in one way or another – 10,000 delegates, 8,000 non-governmental organisations (NGOs) and representatives of 'ordinary' society, 4,000 press, and 1,000 world leaders giving addresses – this was by far the biggest international gathering ever held in Africa.

Johannesburg had no dramatic outcomes. There were no new treaties but lots of new targets, including: to halve the proportion of people without access to basic sanitation by 2015; to use and produce chemicals by 2020 in ways that do not lead to significant adverse effects on human health and the environment; to maintain or restore depleted fish stocks to levels that can produce the maximum sustainable yield on an urgent basis and where possible by 2015; and to achieve by 2010 a significant reduction in the rate of loss of biological diversity. One of the key differences between Johannesburg and Rio was the different approaches taken to the problems raised. The 2002 meeting involved far more dialogue with the groups originally identified in Agenda 21: business, farmers, indigenous peoples, industry, local authorities, NGOs, scientists, technologists, trade unions, 'women', 'workers' and young people. And commitments were made not just by governments but also by a host of others – NGOs, intergovernmental organisations and businesses. Over 300 voluntary initiatives were launched.

US secretary of state Colin Powell declared the summit a 'successful effort'. However, not all participants were equally as pleased. Venezuelan president Hugo Chávez (the man who at the time was selling petrol at 12 cents a gallon to Venezuelans), representing the views of 132 lower-income countries, agreed it was successful but was equally clear that it was still not enough, and in fact some of the steps taken could have been considered as steps backward. Many of the NGOs also thought that the summit did not go far enough, particularly when it came to target setting for increasing the use and development of renewable energies. While disappointed with some of the outcomes, the president of the World Resources Institute, Jonathan Lash, commented positively: 'This summit will be remembered not for the treaties, the commitments, or the declarations it produced, but for the first stirrings of a new way of governing the global commons – the beginnings of a shift from the stiff formal waltz of traditional diplomacy to the jazzier dance of improvisational solution-oriented partnerships that may include non-government organisations, willing governments and other stakeholders.'

Millennium Assessment

To start the first year of a new century and a new millennium on a good footing, the UN produced Resolution 55/2, the United Nations Millennium Declaration, which contained eight specific goals: to eradicate extreme poverty and hunger; to achieve universal primary education; to promote gender equality and empower women; to reduce child mortality; to improve maternal health; to combat HIV/AIDS, malaria and other diseases; to ensure environmental sustainability; and to develop a global partnership for development. They are referred to collectively as the UN Millennium Development Goals. All 191 UN member states pledged to meet these goals by 2015. The resolution contained a section on respect for nature which says, 'The current unsustainable patterns of production and consumption must be changed in the interest of our future welfare and that of our descendants.' It also specifically states that the ongoing and worsening degradation of ecosystem services is a stumbling block to achieving these Millennium Development Goals.

Against this backdrop, in 2001 the UN initiated a massive study which brought together well over one thousand experts, representing ninety-five different countries, to: produce a global inventory of the state of our ecosystems; quantify the effect that human activities are having on them; and make suggestions for the future. The results were published in the summer of 2005 in *The Millennium Ecosystem Assessment Synthesis Report*. It states that 'any progress achieved in addressing the goals of poverty and hunger eradication, improved health, and environmental protection is unlikely to be sustained if most of the ecosystem services on which humanity relies continue to be degraded.' Comprehensive in both its coverage and depth, one of the report's key takeaways is that continuing degradation of fifteen of the twenty-four ecosystem services the Earth provides is increasing the likelihood of abrupt changes that will have a serious and negative impact on our own well-being. They give examples of sudden changes in water quality, shifts in climate, fisheries collapse and the emergence of new diseases.

Aichi (2010) and 'Pathway for Humanity' (2015)

Strategic plan for biodiversity and Aichi biodiversity targets

In 2010 the CBD put forward *The Strategic Plan for Biodiversity*, which had at its centre the twenty Aichi biodiversity targets (Table 4). These targets are clustered under five key goals: a) address the underlying drivers of biodiversity loss; b) reduce direct pressures; c) improve the status of biodiversity; d) enhance the benefits of biodiversity for all; and e) enhance the implementation. Also within the plan was the 2050 Vision for Biodiversity, which aimed to ensure that 'by 2050, biodiversity is valued, conserved, restored and wisely used, maintaining ecosystem services, sustaining a healthy planet and delivering benefits essential for all people.'

Also in 2010 the CBD adopted a supplementary agreement to the original convention. Since 1992 there had been concern about how the fair and equitable sharing of benefits arising out of the utilisation of genetic diversity did not seem to play a large part in the CBD's thinking about biodiversity, even though it was prominent in the Rio declaration as one of three key objectives of the convention. This supplement was a legal framework entitled *The Nagoya Protocol on Access to Genetic Resources and the Fair and Equitable Sharing of Benefits Arising from their Utilization to the Convention on Biological Diversity*.

'Pathway for Humanity': UN Sustainable Development Goals (2015)

Building on the Millennium Development Goals, UN Resolution 70/1, Transforming our World: The 2030 Agenda for Sustainable

Table 4 Summary of progress on Aichi biodiversity targets

Goal	Target element (abbreviated)	Progress towards Aichi targets		
		Poor	Moderate	Good
A) Address the underlying drivers	1.1 Awareness of biodiversity		■	
	1.2 Awareness of steps to conserve		■	
	2.1 Biodiversity integrated into poverty reduction		■	
	2.2 Biodiversity integrated into planning		■	
	2.3 Biodiversity integrated into accounting	■		
	2.4 Biodiversity integrated into reporting	■		
	3.1 Harmful subsidies eliminated and reformed	■		
	3.2 Positive incentives developed and implemented	■		
	4.1 Sustainable production and consumption	■		
	4.2 Use within safe ecological limits	■		
B) Reduce direct pressures	5.1 Habitat loss at least halved	■		
	5.2 Degradation and fragmentation reduced	■		
	6.1 Fish stocks harvested sustainably	■		
	6.2 Recovery plans for depleted species		Unknown	
	6.3 Fisheries have no adverse impact	■		
	7.1 Agriculture is sustainable	■		
	7.2 Aquaculture is sustainable	■		
	7.3 Forestry is sustainable		■	
	8.1 Pollution not detrimental	■		
	8.2 Excess nutrients not detrimental	■		

Table 4 Summary of progress on Aichi biodiversity targets (*continued*)

B) Reduce direct pressures	9.1 Invasive alien species prioritised			
	9.2 Invasive alien pathways prioritised		Unknown	
	9.3 Invasive species controlled or eradicated			
	9.4 Invasive introduction pathways managed			
	10.1 Pressure on coral reefs minimised			
	10.2 Pressures on vulnerable ecosystems minimised			
C) Improve biodiversity status	11.1 Ten per cent of marine areas conserved			
	11.2 Seventeen per cent of terrestrial areas conserved			
	11.3 Areas of importance conserved			
	11.4 Protected areas, ecologically representative			
	11.5 Protected areas, effectively and equitably managed			
	11.6 Protected areas, well connected and integrated			
	12.1 Extinctions prevented			
	12.2 Conservation status of threatened species improved			
	13.1 Genetic diversity of cultivated plants maintained			
	13.2 Genetic diversity of farmed animals maintained			
	13.3 Genetic diversity of wild relatives maintained			
	13.4 Genetic diversity of valuable species maintained		Unknown	
	13.5 Genetic erosion minimised			

Table 4 Summary of progress on Aichi biodiversity targets (*continued*)

D) Enhance benefits to all	14.1 Ecosystems providing services restored and safeguarded			
	14.2 Taking account of women, indigenous peoples and local communities (IPLCs), and other groups		Unknown	
	15.1 Ecosystem resilience enhanced		Unknown	
	15.2 Fifteen per cent of degraded ecosystems restored		Unknown	
	16.1 Nagoya Protocol in force			
	16.2 Nagoya Protocol operational			
E) Enhance implementation	17.1 National biodiversity strategies and action plans (NBSAPs) developed and updated			
	17.2 NBSAPs adopted as policy instruments			
	17.3 NBSAPs implemented			
	18.1 Indigenous and local knowledge (ILK) and customary use respected			
	18.2 ILK and customary use integrated		Unknown	
	18.3 IPLCs participate effectively		Unknown	
	19.1 Biodiversity science improved and shared			
	19.2 Biodiversity science applied		Unknown	
	20.1 Financial resources for Strategic Plan increased			

'Good' = substantial positive trends at a global scale relating to most aspects of the element. 'Moderate' = the overall global trend is positive, but insubstantial or insufficient, or there may be substantial positive trends for some aspects of the element, but little or no progress for others; or the trends are positive in some geographical regions, but not in others. 'Poor' = little or no progress towards the element or movement away from it; or despite local, national or case-specific successes and positive trends for some aspects, the overall global trend shows little or negative progress or there is insufficient information to score progress. (Source: *Global Assessment Report on Biodiversity and Ecosystem Services* (2019).)

Development, recognised that 'eradicating extreme poverty is the greatest global challenge and an indispensable requirement for sustainable development.' This caused the UN General Assembly in 2015 to action seventeen 'Global Goals' (or the Sustainable Development Goals) which they aimed to achieve by 2030 (and sometimes before). These were integrated and indivisible goals which, while they included ending extreme poverty, acknowledged the interconnectedness of life by also focusing on building sustainable cities, giving people better healthcare, achieving equality for women, and of course biodiversity. There are over a hundred specific targets. Only Targets 14 (Life below water) and 15 (Life on land) are explicitly about biodiversity, but in reality, because these genuinely serve integrated and indivisible goals, what happens to biodiversity does impact each of the remaining goals and vice versa.

Global Assessment Report on Biodiversity and Ecosystem Services (2019)

The *Global Assessment Report on Biodiversity and Ecosystem Services* is the first critical assessment of biodiversity since the release of the *Millennium Ecosystem Assessment* in 2005 and the first carried out by an intergovernmental body. There are four main take-home messages, each backed with an evidence base. They are that: a) biodiversity and the services it provides are deteriorating worldwide; b) the rate of deterioration has accelerated over the past fifty years, at a rate unprecedented in human history; c) targets for maintaining biodiversity and achieving sustainability, such as the Aichi biodiversity targets and the 2030 Agenda for Sustainable Development, cannot be met by current trajectories and goals for 2030 – they will only be met by 'transformative changes' (defined as 'fundamental, system-wide reorganisation across technological, economic and social factors, including paradigms, goals and values') across the board, and it is the poorest who will suffer most; d) maintaining biodiversity and other societal goals are still achievable if there is this transformative change.

As the basis for point c), the report critically evaluated progress made towards achieving the Aichi biodiversity targets and the UN Sustainable Development Goals.

Aichi

Some of the Aichi biodiversity targets are partially achieved. The *Global Assessment* concludes that good progress has been made towards four out of twenty. 'Good' means there are substantial positive trends at a global scale. But the devil is in the detail as sadly only progress in about one in ten components was graded as 'good'.

Target 11: Increasing protected areas

The main success story is undoubtedly Target 11, where two out of six components received a 'good' in terms of progress. This is because in the last decade the amount of protected area has increased by 2.3% on land to a total of 15% globally (this includes fresh water), and by 5.4% in the ocean to a total of 7% globally. The original target for land was 10% conserved, and for the ocean 17%. All of this sounds quite impressive, and in so many ways it is. But it should be remembered that there are also significant problems with the present scheme of protecting areas, which takes the shine off just a little.

Firstly, most of the areas allocated for protection are too small. True, the numbers increase each year, but the average size of the area is getting smaller. Setting up corridors, conduits between reserves, would help, but this is rarely done.

Not only are the areas set aside too small but they tend to be sections of land with low economic value, and rarely are thought through in terms of the patterns of natural occurrence of the animals and plants there. And while areas may be designated as national parks or wilderness areas, in few cases does this mean that they are actually afforded any protection. So in practice many designated areas turn out not to be protected areas at all. These are referred to as paper parks. For instance, when it was set up in January 2000, the Port Honduras Marine Reserve (a marine protected area or MPA in Belize, consisting of 133 islands) was in effect a paper park; There were no buoys marking the extent of the park; the large area (837 square kilometres) was difficult to patrol; and the NGO which had the responsibility for management was small with very limited resources. As a result, poaching by national and foreign fishermen went on unabated. However, the next year, with the aid of a reasonable grant, the NGO built a station in the middle of the reserve that improved their ability to police the area, and also worked with local fishermen to develop economic alternatives to commercial fishing, such as fly fishing, kayaking, and diving and other ecotourism developments. There are now grounds for referring to this park as a marine *protected* area. But this is an exception. Currently 80–90% of marine protected areas are not protected at all.

Finally, as recognised in the CBD, but rarely acted upon, it is often not enough to set up protected areas without also protecting the wider area these protected zones are part of. For example, there are nature reserves set aside for the African elephant, but unfortunately four out of five elephants don't actually live on a reserve. Our commitment to operating such national parks is perhaps best illustrated by the fact that we spend a mere US$6 million on protected areas worldwide, one tenth of what we spend on dieting methods.

Looking forward, the Campaign for Nature, a partnership of the Wyss Campaign for Nature and the National Geographic Society, is calling on policymakers to commit to a science-driven, ambitious new deal for nature that will carry a commitment to protect at least 30% of the planet by 2030. This is to be introduced at the 15th meeting of the Conference of the Parties to the Convention on Biological Diversity in Kunming, China, in 2021.

Goal 16: Nagoya protocol in force

Graded 'good' is the fact that *The Nagoya Protocol*, mainly dealing with genetic resources, is now in force. Adopted on 29 October 2010 in Nagoya, Aichi province, Japan, the protocol came into force almost exactly four years later on 12 October 2014. However, at this time, progress to making it operational is only graded 'moderate'. The Nagoya protocol has turned out to be controversial, with some scientists worried that it will impede research and collaboration.

However, for about one third of the Aichi goals it was acknowledged that there was little or no progress.

Sustainability goals

The *Global Assessment* points out that to meet the UN Sustainable Development Goals, some of the targets set in 2010 need to be reframed and made more effective by taking into account the impacts of climate change. The *Global Assessment* also highlights the fact that current reductions in biodiversity will undermine progress towards thirty-five of forty-four of assessed targets in the goals related to poverty, health, water, cities, climate, ocean and land, as well as the 2050 Vision for Biodiversity.

No room for the individual?

It is widely accepted that individuals can have a great influence on conserving biodiversity and the environment at the local level. But the global problems, as we've seen, seem insurmountable even for governments and intergovernmental committees. What possible influence could an individual have on a global scale?

Wangari Muta Maathai was born in Nyeri, Kenya, in 1940. She was the first woman in East and Central Africa to earn a doctoral degree. She became chair of the Department of Veterinary Anatomy at the University of Nairobi in 1976, the first woman to hold such a position in the region. She was active in the National Council of Women of Kenya between 1976 and 1987, being its chairperson from 1981 onward. It was during this time that she hit upon the idea of planting trees with women's groups in order to both conserve the environment and to improve their quality of life. During her speech accepting the Nobel Peace Prize in 2004, she said, 'My inspiration partly comes from my childhood experiences and observations of nature in rural Kenya... As I was growing up, I witnessed forests being cleared and replaced by commercial plantations, which destroyed local biodiversity and the capacity of the forests to conserve water.'

Through the organisation she formed, the Green Belt Movement, she assisted women in planting more than 20 million trees on their farms and in school grounds and church compounds. The organisation expanded into the Pan African Green Belt Network in 1986 and similar tree-planting schemes were extended to countries such as Ethiopia, Lesotho, Malawi, Tanzania, Uganda and Zimbabwe. She played a major role in seeking the cancellation of African debt, through the Jubilee 2000 Coalition, and her hands-on protests against land grabbing and deforestation earned her a high international profile. By 2000 she was internationally recognised for her campaigning for democracy, human rights and environmental conservation.

In December 2002, Professor Maathai was elected to parliament with an overwhelming 98% of the vote and was appointed assistant minister for environment, natural resources and wildlife.

In October 2004 she was awarded the Nobel Peace Prize 'for her contribution to sustainable development, democracy and peace'. Not everyone was pleased with the choice. One politician raged, 'You don't give the Nobel chemistry prize to a professor in economics. A peace prize should honour peace, not the environment.' Also, her controversial (and, if rightly reported, wrong) views on the origin and spread of AIDS caused a bit of a stir. However, in many ways her nomination underlines the point made in this and in the previous chapter – that you cannot really separate concern for biodiversity from concern for the environment generally, and from the multitude of other social and political challenges, including poverty, that seem to go hand in hand. This can even be seen in her acceptance speech, where she said, 'So, together, we have planted over 30 million trees that provide fuel, food, shelter and income to support their children's education and household needs. The activity also creates employment and improves soils and watersheds.'

Both Maathai and her Green Belt Movement have received numerous other awards. She was listed on UNEP's Global 500 Hall of Fame and named one of the hundred heroines of the world. In June 1997, she was selected by *Earth Times* as one of 100 persons in the world who have made a difference in environmental conservation.

Maathai concluded her Nobel Prize acceptance speech with these words:

> I reflect on my childhood experience when I would visit a stream next to our home to fetch water for my mother. I would drink water straight from the stream. Playing among the arrowroot leaves I tried in vain to pick up the strands of frogs' eggs, believing they were beads. But every time I put my little fingers

> under them, they would break. Later, I saw thousands of tadpoles: black, energetic and wriggling through the clear water against the background of the brown earth. This is the world I inherited from my parents.
>
> Today, over fifty years later, the stream has dried up, women walk long distances for water, which is not always clean, and children will never know what they have lost. The challenge is to restore the home of the tadpoles and give back to our children a world of beauty and wonder.

Can one person make a difference? From Maathai's story the answer must be yes. An exceptional person, admittedly, but one person nevertheless – and that person doesn't have to be right in everything they believe.

Epilogue

There have been great advances made in our thinking about what to conserve and how to conserve, even on a global scale. The end of the last century saw the first global attempt to tackle our biodiversity crisis, given substance in the words of the Convention for Biological Diversity. Unfortunately, as we have seen, our subsequent actions and resolve have often not kept pace with either our enlightened thinking or our words. And the pace of biodiversity loss continues to accelerate. The political and moral will to take the conservation of biodiversity seriously is still lacking. Despite mounting scientific evidence, much of it presented in previous chapters, there still seems an unwillingness to face the problem of our biodiversity crisis. One academic wrote: 'there is unfortunately no precedent for 5 [currently 7.6] billion human beings suddenly sharing an enlightened vision of the future.'

But now press pause. That was before the beginning of 2020. It is at least possible that the COVID-19 pandemic, and our collective reaction to it, may encourage or force us to rethink and reimagine much that we have up until now ignored. This must surely include the realisation that biodiversity and human life are inextricable. If you damage one, you most certainly damage the other. COVID-19 knocked biodiversity (and climate change) off the agenda. The UN Convention on Biological Diversity published their *Global Biodiversity Outlook 5* in September 2020, the final report card on progress made on the Aichi biodiversity targets. None of the major aims were achieved in full, although, as we saw above, there were some limited successes. Published in the midst of what can only be described as a time of dramatic and unexpected transformative change sparked by COVID-19, this new UN report reiterated much of what was published in 2019's *Global Assessment Report on Biodiversity and Ecosystem Services*. But now, in September 2020, it has a new feel, and a renewed urgency. The report's recommended eight transitions away from 'business as usual' contain much that is familiar from previous similar documents. However, they feel new and maybe even possible now. The UN report shows that though the successes of the last ten years are, to say the least, limited, they are possible. Change is possible. One of the completely new elements in the UN report is that COVID-19 has demonstrated our own vulnerability and the vulnerability of the biodiversity we need just to survive. The present context transforms how the document reads, and hopefully how it is received having been published at the beginning of 2020. What now comes across strongly is that when we damage biodiversity, we damage ourselves – but there is a way out. We need transformative change. We've seen and experienced for ourselves in 2020 that such change is possible. If correct, what must be one of the most disappointing UN environment reports, surprisingly, and for the first time, could give tangible hope, as it outlines a path that we know is not just needful but possible.

There are five features of current twenty-first-century life that may well play a key role in recognising that biodiversity is central to any life we are to have on Earth: 1) our continuing response to COVID-19; 2) realising and 3) acting upon the interconnectedness of biodiversity; 4) preparing for almost inevitable future pandemics; and 5) global inequality and poverty. Nearly all of the predicted responses and forecasts in this book were made pre-COVID-19. At this moment in time it is difficult to know how to revise the predictions of what will happen to biodiversity. It could be that COVID-19 acts as a wake-up call to act to maintain 'our common home'. Equally it could throw the world into playing 'catch-up', with the result that we may even exceed a business-as-usual trajectory, to the detriment of biodiversity and ourselves. Or it could be something in between. At the moment I write this it is too close to call.

We can afford to maintain biodiversity. But only after we accept that we cannot afford not to.

8

No one is too small to make a difference

> Some would say we are wasting lesson time [by striking]; we say we are changing the world. So that when we are older, we will be able to look our children in the eyes and say we did everything we could back then. Because that is our moral duty, and we will never stop doing that. We will never stop fighting for the living planet and for a safe future – for our future… We have been told so many times that there's no point in doing this… that we can't have an impact and make a difference. But I think we have proven that to be wrong by now… We are the change, and change is coming.
>
> Greta Thunberg, sixteen-year-old environmental activist, in an address to half a million people at Week for Future, Montreal, Canada, 27 September 2019

I remember one particular ordinary evening when Kevin Gaston and I were still researching information for our textbook *Biodiversity: An Introduction*. I was poring over the latest tranche of relevant scientific papers. Totally engrossed, and with my critical hat on, I was checking out, calculating and estimating extinction rates, gauging how rigorous, how robust, the information on known extinctions was. Almost involuntarily, I found myself putting my notes and papers down and going quietly to where my children were sleeping peacefully. I became aware of a deep ache, almost like a physical pain, growing within me. I find that tears come to me more readily as I grow older. What sort of world would they grow up into? At best my generation was passing on to them a world considerably poorer in terms of the diversity of living things. That ring of living beauty (remember the volleyball on Mission Beach?) was indeed a thin and fragile one. The ache, and a determination to do something about it, has stayed with me ever since. That's what drove me to write this book.

COVID-19 pandemic or not, the question is not how can we avert our biodiversity crisis, but what sort of world do we want to live in? Rachel Carson, an eminent marine biologist from the first half of the twentieth century, wrote an influential book called *Silent Spring*. In it she pictured a world devastated by the indiscriminate use of pesticides like DDT. The world she wrote of was a world devoid of birds singing. The image was so powerful, and the book so moving, that it changed the way a whole generation thought about pollution. Faced with the question of what sort of world they wanted to live in, the readers of Carson's book, and the world in general, opted for one where birds did sing.

So, for those of us living now at the beginning of this new millennium, again the question is 'what sort of world do you want to live in?' One where diseases like COVID-19 become more prevalent because we have so weakened the natural infrastructure of biodiversity? One where economic and physical wars are increasingly fought over water, energy and access to other resources, where starvation and poverty are even more prevalent because we have destroyed much of the biodiversity that gives us the ability to exist, never mind live? A world where only the increasingly mega-rich can pay the enormous amounts of money necessary to secure dwindling services that once were free; a world devoid, stripped, of all natural beauty and capital, where the variety of life is something that only exists in books or stored in electronic format in 'The Cloud'? A world where the predominant 'natural' colour is grey and where mere existence, just getting by, just hanging on a little longer, becomes an even more acceptable and inevitable option over *living*?

An interest in biodiversity is not just 'a passion for wildlife' or a 'commitment to favourite conservation issues', although it may involve those things. It's not just a scientific discipline, although it can be studied as such, as we have done – and much pleasure can be derived from that study alone. Biodiversity is intimately and inextricably tied up with our existence, our lives and our lifestyles. It doesn't matter whether you've got 'green tendencies' or you're a hard-nosed business person, whether you think that saving the lesser spotted weevil from extinction should be an all-consuming passion or your days are spent merely getting by, surviving, trying to make sure your family is okay. The natural capital and resilience that is biodiversity gives us life, gives us options. Everything that is the basis of our society is provided by nature. Biodiversity is resilient to change and pressure, but only up to a point. And there are now enough indications that that point has been passed. The destruction and degradation of biodiversity is the beginning of our own destruction, with the poorest people suffering first and most.

For those who, like me, have belief in a God who created all this biodiversity, and to whom I believe this biodiversity belongs, the realisation that we have been poor stewards should fill us with deep sadness, but our faith should drive a willingness to do something, to get involved. The history of ordinary people is a history of suffering and injustice. I believe Jesus, driven by compassion, gave everything to change that situation and expected his followers to do the same. Our biodiversity crisis is already bringing suffering to the poorest and the most marginalised, and this will only get worse. My understanding is that while 'environmental concerns' do not seem to be high up the priority list of some of the Christians with whom, at least on paper, I share a similar belief, they do rank high in the purposes of our God. And this demands some response: 'What sort of world does God want us to live in?' I know it is not the one I am passing on to my children.

The bottom line is, no matter who you are, biodiversity matters. It matters to you, your family, everyone you know, and everyone you do not know and will never meet, both at the highest levels and at the most fundamental levels of existence. And that's why it's worth knowing about. We cannot avert our biodiversity crisis. But we can do something. We should not allow the question 'What sort of world do we want to live in?' to go unanswered or, even worse, unconsidered. At the end of the twentieth century, Mother Teresa said that poverty is so much more than not having money. It is a dreadful thing to waste our lives and the lives of all the living things around us by default.

So what do we do? First and foremost, I suggest you believe that you and I as individuals can make a difference. Many young people inspired by Greta Thunberg and their own sense of justice have begun to catch on to that. It is, in my view, one of the most exciting developments of the new millennium.

We can, and should, lobby our politicians. We could get involved in local politics. We can think of innovative ways of conserving biodiversity at home, locally, nationally and even globally (see some suggestions in the further reading section). We can get involved with non-governmental organisations that are fighting for sustainability, equality and environmental change, and against poverty and the greed that drives much of our biodiversity crisis. We must change the way we live. Most of all, you and I must believe that we can change things even in the midst of not only our biodiversity crisis, but our climate crisis, our COVID-19 crisis, and whatever other crises are waiting for us round the corner. As someone has said, if we, the generation that faces this next century, do not do the impossible, we will be faced with the unthinkable. When we damage biodiversity, we damage ourselves; the poor feel it first, but even the rich will not be immune. And I do not see why my children, our children, anyone's child, should be faced with that.

Going further: Suggestions for wider reading

All website links were accurate at the time of writing (November 2020).

1 The pandemic of wounded biodiversity

Casetta E, Marques da Silva J, Vecchi D (editors), 2019. *From Assessing to Conserving Biodiversity: Conceptual and Practical Challenges*. Springer.

Lanzerath D, Friele M, 2014. *Concepts and Values in Biodiversity*. Routledge.

Levin SA (editor), 2013. *Encyclopedia of Biodiversity* (5 volumes) (2nd edition). Academic Press. (Updated and expanded edition, consists of 335 separate articles and covers the dimensions of diversity, an introduction to the services biodiversity provides, and measures to protect it.)

MacKenzie D, 2020. *Covid-19: The Pandemic that Never Should Have Happened and How to Stop the Next One*. The Bridge Street Press.

Magurran AE and McGill BJ (editors), 2011. *Biological Diversity: Frontiers in Measurement and Assessment*. Oxford University Press.

Strauss R, 2008. *Tree of Life: The Incredible Biodiversity of Life on Earth*. A&C Black Publishers Ltd.

Wilson EO (editor), 1988. *Biodiversity*. John Wiley & Sons. (Out of date but still worth reading, the whole of this landmark book can be read online at http://books.nap.edu/books/0309037395/html/index.html.)

Wilson EO, 2021. *Biophilia, The Diversity of Life, Naturalist*. Library of America. (Rerelease of three Wilson classics.)

2 Teeming boisterous life

There are a large number of people and organisations that are considerably more knowledgeable about the biodiversity in and around La Jolla than I am. So if you want to find out more about the biodiversity of this fascinating region, Kelly Stewart is a marine biologist with the Ocean Foundation (https://oceanfdn.org) who writes about the flora and fauna of La Jolla. Here is an example of one of her pieces: https://www.lajollalight.com/news/local-news/sdljl-natural-la-jolla-lizards-2016jun08-story.html. And she gives her email at the end of the piece. Or check out the websites of the San Diego Natural History Museum (https://www.sdnhm.org), the Center for Marine Biodiversity and Conservation at the Scripps Institution of Oceanography (https://escholarship.org/uc/sio_cmbc), the San Diego Coastkeeper (https://www.sdcoastkeeper.org/blog/category/marine-conservation), and the Birch Aquarium at Scripps (https://aquarium.ucsd.edu).

iNaturalist (www.inaturalist.org) is an online social network to help people identify (crowdsourced species identification system) and then to share (organism occurrence recording tool) biodiversity information. Its originators say that its primary goal is to connect people to nature. Used with care it can be an enjoyable way to identify and map biodiversity.

Amyes SGB, 2013. *Bacteria: A Very Short Introduction*. Oxford University Press.

Anderson JM, Anderson HM, Cleal CJ, 2007. *Brief History of the Gymnosperms: Classification, Biodiversity, Phytogeography and Ecology*. SANBI Publishing.

Antonelli A *et al.*, 2020. *State of the World's Plants and Fungi 2020*. Royal Botanic Gardens, Kew. (https://doi.org/10.34885/172)

Arndt I, 2014. *Animal Architecture*. Abrams.

Attenborough D, 2018. *Life on Earth*. William Collins.

Crawford DH, 2018. *Viruses: A Very Short Introduction*. Oxford University Press.

Crous PW *et al.*, 2009. *Fungal Biodiversity*. CBS Laboratory Manual series. Centraalbureau voor Schimmelcultures.

Dipper F, Dando M, 2016. *The Marine World: A Natural History of Ocean Life*. Wild Nature Press.

Engel MS, 2018. *Innumerable Insects: The Story of the Most Diverse and Myriad Animals on Earth*. Sterling.

Giribet G, Edgecombe GD, 2020. *The Invertebrate Tree of Life*. Princeton University Press.

Helfman GS *et al.*, 2009. *The Diversity of Fishes: Biology, Evolution, and Ecology*. Wiley-Blackwell.

Holland P, 2011. *The Animal Kingdom: A Very Short Introduction*. Oxford University Press.

Linzey DW, 2020. *Vertebrate Biology* (3rd edition). Johns Hopkins University Press.

Money NP, 2016. *Fungi: A Very Short Introduction*. Oxford University Press.

Petersen JH, 2013. *The Kingdom of Fungi*. Princeton University Press.

Pough FH, Janis CM, 2018. *Vertebrate Life* (10th edition). Oxford University Press.

Poulin R, Morand S, 2014. *Parasite Biodiversity*. Smithsonian Institution Scholarly Press.

Scales H, 2016. *Spirals in Time: The Secret Life and Curious Afterlife of Seashells*. Bloomsbury.

Sheldrake M, 2020. *Entangled Life: How Fungi Make Our Worlds, Change Our Minds, and Shape Our Futures*. Bodley Head.

Steinbeck J, Ricketts EF, 2009. *Sea of Cortez: A Leisurely Journal of Travel and Research*. Penguin.

Strauss R, 2008. *Tree of Life: The Incredible Biodiversity of Life on Earth*. A&C Black.

Sverdrup-Thygeson A, 2018. *Extraordinary Insects: Weird, Wonderful, Indispensable: The Ones Who Run Our World*. Mudlark.

Taberlet P *et al.*, 2018. *Environmental DNA: For Biodiversity Research and Monitoring*. Oxford University Press.

Tudge C, 2002. *The Variety of Life: A Survey and a Celebration of All the Creatures that Have Ever Lived*. Oxford University Press.

Walker T, 2012. *Plants: A Very Short Introduction*. Oxford University Press.

Youatt R, 2015. *Counting Species: Biodiversity in Global Environmental Politics*. University of Minnesota Press.

3 Where on Earth is biodiversity?

Lots of good stuff on biodiversity hotspots can be found at https://www.biodiversityhotspots.org/xp/Hotspots, much of which is an elaboration of Norman Myers's original scheme (see p. 71).

Cox CB, Moore PD, Ladle RJ, 2016. *Biogeography: An Ecological and Evolutionary Approach* (9th edition). Wiley-Blackwell.

David B, Saucède T, 2015. *Biodiversity of the Southern Ocean*. ISTE Press.

Dudgeon D, 2020. *Freshwater Biodiversity: Status, Threats and Conservation*. Cambridge University Press.

Erwin DH, 2015. *Extinction: How Life on Earth Nearly Ended 250 Million Years Ago*. Princeton Science Library. Princeton University Press.

Franklin J, 2010. *Mapping Species Distributions: Spatial Inference and Prediction*. Cambridge University Press.

Giribet G, Edgecombe GD, 2020. *The Invertebrate Tree of Life*. Princeton University Press.

Groombridge B, Jenkins MD, Jenkins M, 2002. *World Atlas of Biodiversity: Earth's Living Resources in the 21st Century*. University of California Press.

Hoom C, Perrigo A, Antonelli A (editors), 2018. *Mountains, Climate and Biodiversity*. Wiley-Blackwell.

Hubbell SP, 2011. *The Unified Neutral Theory of Biodiversity and Biogeography*. Princeton University Press.

Lomolino MV, Riddle BR, Whittaker RJ, 2016. *Biogeography: Biological Diversity Across Space and Time* (5th edition). Oxford University Press.

Lomolino MV, 2020. *Biogeography: A Very Short Introduction*. Oxford University Press.

Matthews TJ, Triantis KA, Whittaker RJ (editors), 2020. *The Species–Area Relationship: Theory and Application*. Cambridge University Press.

Mittermeier RA, Gil PR, Hoffman M, 2005. *Hotspots Revisited: Earth's Biologically Richest and Most Endangered Terrestrial Ecoregions*. University of Chicago Press.

Rull V, 2020. *Quaternary Ecology, Evolution, and Biogeography*. Academic Press.

Wang Y (editor), 2020. *Terrestrial Ecosystems and Biodiversity*. CRC Press.
Worm B, Tittensor DP, 2018. *A Theory of Global Biodiversity*. Monographs in Population Biology (vol. 60). Princeton University Press.
Zachos FE, Habel JC (editors), 2011. *Biodiversity Hotspots: Distribution and Protection of Conservation Priority Areas*. Springer.

4 A world that was old when we came into it

The Living Planet Reports are published every two years and you can download the 2020 report at https://livingplanet.panda.org/en-us.

Benton MJ, 2015. *When Life Nearly Died: The Greatest Mass Extinction of All Time*. Thames and Hudson Ltd.
Benton MJ, 2019. *Cowen's History of Life*. Wiley-Blackwell.
Benton MJ, Harper DAT, 2020. *Introduction to Paleobiology and the Fossil Record* (2nd edition). John Wiley & Sons.
Bottjer DJ, 2016. *Paleoecology: Past, Present and Future*. Wiley-Blackwell.
Bromham L, Cardillo M, 2019. *Origins of Biodiversity: An Introduction to Macroevolution and Macroecology*. Oxford University Press.
Dawson A, 2016. *Extinction: A Radical History*. OR Books.
Deamer DW, 2020. *Origin of Life: What Everyone Needs to Know*. Oxford University Press.
Erwin DH, Valentine JW, 2013. *The Cambrian Explosion: The Construction of Animal Biodiversity*. Roberts and Company Publishers.
Hazen RM, 2013. *The Story of Earth: The First 4.5 Billion Years, from Stardust to Living Planet*. Penguin Random House US.
Knoll AH, 2015. *Life on a Young Planet: The First Three Billion Years of Evolution on Earth*. Princeton Science Library. Princeton University Press.
Lewis S, Maslin MA, 2018. *The Human Planet: How We Created the Anthropocene*. Pelican Books.
Luisi PL, 2019. *The Emergence of Life: From Chemical Origins to Synthetic Biology*. Cambridge University Press.

MacLeod N, 2013. *The Great Extinctions: What Causes Them and How They Shape Life.* The Natural History Museum.
McGhee Jr GR, 2019. *Convergent Evolution on Earth: Lessons for the Search for Extraterrestrial Life.* The Vienna Series in Theoretical Biology. MIT Press.
Wackernagel M, Beyers B, 2019. *Ecological Footprint: Managing Our Biocapacity Budget.* New Society Publishers.
Xian-Guang H *et al.*, 2017. *The Cambrian Fossils of Chengjiang, China: The Flowering of Early Animal Life.* Wiley-Blackwell.

5 Swept away and changed

The World Wildlife Fund (US and Canada) or the World Wide Fund for Nature (rest of the world) (WWF) works to preserve wilderness areas and reduce the impact of humans on the environment. You can find them online at https://www.worldwildlife.org.

The Intergovernmental Panel on Climate Change, who provide the most authoritative information on climate change, can be found online at http://www.ipcc.ch.

Anthony L, 2018. *The Aliens Among Us: How Invasive Species Are Transforming the Planet – and Ourselves.* Yale University Press.
FAO, 2019. *The State of The World's Biodiversity for Food and Agriculture.* Bélanger J, Pilling D (editors). FAO Commission on Genetic Resources for Food and Agriculture Assessments. (http://www.fao.org/3/CA3129EN/CA3129EN.pdf)
FAO, 2020. *The State of World Fisheries and Aquaculture 2020: Sustainability in Action.* FAO. (http://www.fao.org/3/ca9229en/ca9229en.pdf)
FAO, IFAD, UNICEF, WFP, WHO, 2020. *The State of Food Security and Nutrition in the World 2020: Transforming Food Systems for Affordable Healthy Diets.* FAO. (https://doi.org/10.4060/ca9692en)
FAO and UNEP, 2020. *The State of the World's Forests 2020: Forests, Biodiversity and People.* FAO and UNEP. (https://doi.org/10.4060/ca8642en)

Foreman D, Carroll L, 2015. *Man Swarm: How Overpopulation Is Killing the Wild World* (2nd edition). LiveTrue Books.

Goldemberg J, Ferguson CD, Prud'homme A, 2015. *The World's Energy Supply: What Everyone Needs to Know*. Oxford University Press. (Compiles *Energy: What Everyone Needs to Know*, *Nuclear Energy: What Everyone Needs to Know*, and *Hydrofracking: What Everyone Needs to Know*.)

Kellert SR, 2018. *Nature by Design: The Practice of Biophilic Design*. Yale University Press.

Lange G-M, Woden Q, Carey K (editors), 2018. *The Changing Wealth of Nations 2018: Building a Sustainable Future*. World Bank Group.

Lovejoy TE, Hannah L, 2019. *Biodiversity and Climate Change: Transforming the Biosphere*. Yale University Press.

Luisi PL, 2019. *The Emergence of Life: From Chemical Origins to Synthetic Biology*. Cambridge University Press.

Maslin M, 2014. *Climate Change: A Very Short Introduction*. Oxford University Press.

Morris J, 2016. *Sustainable Control of Invasive Plants: Adding Value to Biodiversity Economy for Ecosystem Services*. LAP Lampert.

Probert PK, 2017. *Marine Conservation*. Cambridge University Press.

Reckhaus H-D, 2018. *Why Every Fly Counts: A Documentation About the Value and Endangerment of Insects*. Springer.

Roberts C, 2013. *The Ocean of Life: The Fate of Man and The Sea*. Penguin.

Simberloff D, 2013. *Invasive Species: What Everyone Needs to Know*. Oxford University Press.

Thompson K, 2015. *Where Do Camels Belong? The Story and Science of Invasive Species*. Profile Books.

UN, DESA, 2020. *World Economic Situation and Prospects 2020*. UN and DESA. (https://www.un.org/development/desa/dpad/wp-content/uploads/sites/45/publication/WESP2020_FullReport_web.pdf)

Walker E, Menu S, 2020. *Biomimicry: When Nature Inspires Amazing Inventions*. Triangle Square Press.

Wallace-Wells D, 2019. *The Uninhabitable Earth: A Story of the Future*. Penguin.

Wilson EO, 2017. *Half-Earth: Our Planet's Fight for Life*. Liveright.

Willis KJ (editor), 2017. *State of the World's Plants 2017*. Royal Botanic Gardens, Kew. (https://kew.iro.bl.uk/work/ns/2e0d292a-c3da-49ea-a500-32a4f9aff281)

Willis KJ (editor), 2018. *State of the World's Fungi 2018*. Royal Botanic Gardens, Kew. (https://kew.iro.bl.uk/work/ns/e30de436-455d-410e-8605-8c533a0398ce)

WWF, 2020. *Living Forests Report 2020*. (https://wwf.panda.org/our_work/our_focus/forests_practice/forest_publications_news_and_reports/living_forests_report/)

6 Are the most beautiful things the most useless?

The official website of the Biosphere 2 project can be found at https://biosphere2.org, but it's also worth checking out http://www.bioquest.org/simbio2.html, an interactive site which models the Biosphere 2 project and allows you to rerun the experiment, changing things like the number of people in the closed system and the sorts of organism included.

Abdul-Matin I, 2010. *GreenDeen: What Islam Teaches About Protecting the Planet*. Berrett-Koehler Publishers.

Anderson V, 2019. *Debating Nature's Value: The Concept of 'Natural Capital'*. Palgrave Pivot.

Attfield R, 2018. *Environmental Ethics: A Very Short Introduction*. Oxford University Press.

Bacchus S, 2020. *Environmental Ethics from a Faith-Based Perspective*. FriesenPress.

Bauman W, O'Brien KJ, 2019. *Environmental Ethics and Uncertainty: Wrestling with Wicked Problems*. Routledge.

Brennan A, 2015. *Thinking About Nature: An Investigation of Nature, Values and Ecology*. Routledge.

Case-Winters A, 2016. *Reconstructing a Christian Theology of Nature: Down to Earth*. Routledge.

Conradie EM *et al.* (editors), 2015. *Christian Faith and the Earth: Current Paths and Emerging Horizons in Ecotheology*. Bloomsbury.

Framarin CG, 2014. *Hinduism and Environmental Ethics*. Routledge.

Gade AM, 2019. *Muslim Environmentalisms: Religious and Social Foundations*. Columbia University Press.

Gardiner SM, Thompson A, 2019. *The Oxford Handbook of Environmental Ethics*. Oxford University Press.

Garson J *et al.*, 2016. *The Routledge Handbook of Philosophy of Biodiversity*. Routledge.

Grunewald K, Bastian O, 2015. *Ecosystem Services: Concept, Methods and Case Studies*. Springer.

Gunnell K, Murphy BL, Williams C, 2013. *Designing for Biodiversity: A Technical Guide for New and Existing Buildings*. RIBA.

Jacobs S, Dendoncker N, Keune H (editors), 2013. *Ecosystem Services: Global Issues, Local Practices*. Elsevier.

Jakobsen O, 2019. *Transformative Ecological Economics: Process Philosophy, Ideology and Utopia*. Routledge.

Jax K, 2010. *Ecosystem Functioning*. Cambridge University Press.

Jenkins W, 2013. *Ecologies of Grace: Environmental Ethics and Christian Theology*. Oxford University Press.

Jenkins W, Tucker ME, Grim J (editors), 2016. *Routledge Handbook of Religion and Ecology*. Routledge.

Jones L, 2020. *Losing Eden: Why Our Minds Need the Wild*. Allen Lane.

Lanzerath D, Friele M (editors), 2014. *Concepts and Values in Biodiversity*. Routledge.

Laurent E, 2019. *The New Environmental Economics: Sustainability and Justice*. Polity.

Maier DS, 2012. *What's So Good About Biodiversity? A Call for Better Reasoning About Nature's Value*. Springer.

Marley C, 2015. *Biophilia*. Abrams.

McCord EL, 2012. *The Value of Species*. Yale University Press.

Morand S, Lajaunie C, 2017. *Biodiversity and Health: Linking Life, Ecosystems and Societies*. Elsevier.

Pearce D, Moran D, 2013. *The Economic Value of Biodiversity*. Routledge.

Rolston H, 2020. *A New Environmental Ethics: The Next Millennium for Life on Earth* (2nd edition). Routledge.

Sala E, 2020. *The Nature of Nature: Why We Need the Wild*. National Geographic.

Seaborg D, 2021. *How Life Increases Biodiversity: An Autocatalytic Hypothesis*. CRC Press.

Smith IA, 2018. *The Intrinsic Value of Endangered Species*. Routledge.
Spash CL (editor), 2018. *Routledge Handbook of Ecological Economics: Nature and Society*. Routledge.
Turner JS, 2007. *The Tinkerer's Accomplice: How Design Emerges from Life Itself*. Harvard University Press.
Van Beukering PJH *et al.*, 2013. *Nature's Wealth: The Economics of Ecosystem Services and Poverty*. Cambridge University Press.

7 Our greatest hazard and our only hope?

In terms of the sorts of organisations involved in conservation in its widest sense, it's worth checking the websites for the UN Environment Programme (https://www.unenvironment.org) as well as the joint venture with the World Conservation Monitoring Centre (http://www.unep-wcmc.org). The Organisation for Economic Co-operation and Development (https://www.oecd.org) is an international organisation that, together with governments, policymakers and citizens, seeks to establish evidence-based international standards and find solutions to a range of social, economic and environmental (including biodiversity) problems.

The official website of the Convention on Biological Diversity can be found at https://www.cbd.int, and the text of the convention at https://www.cbd.int/convention/text. The website of the UN Department of Economic and Social Affairs, Division of Sustainable Development is https://sdgs.un.org/, and that of the Johannesburg World Summit on Sustainable Development 2002 can be found at http://www.johannesburgsummit.org. The official site for the 2002 Earth Summit, mainly for stakeholders (beats garlic, I suppose), is at http://www.earthsummit2002.org.

You can find the UN Millennium Development Goals in full at http://www.un.org/millenniumgoals, and an official website that gives a list of indicators to check the extent to which these goals are being achieved is http://millenniumindicators.un.org/unsd/mi/mi_goals.asp. The CBD's strategic plan and Aichi targets are available at https://www.cbd.int/sp.

The Global Assessment Report on Biodiversity and Ecosystem Services is available at https://ipbes.net/global-assessment. IPBES is the Intergovernmental Science-Policy Platform on Biodiversity and Ecosystem Services, an independent body established by ninety-four countries in 2012 to strengthen the science-policy interface for biodiversity and ecosystem services for the conservation and sustainable use of biodiversity, long-term human well-being, and sustainable development. It has affiliations with, but is not itself a body of, the United Nations.

The Campaign for Nature can be found online at https://www.campaignfornature.org.

For more details on Wangari Muta Maathai, see her pages at the Nobel Prize website, https://www.nobelprize.org/prizes/peace/2004/maathai/biographical/.

Attenborough D, 2020. *A Life on Our Planet: My Witness Statement and a Vision for the Future*. Ebury Press.

Barnes S, 2018. *Rewild Yourself: 23 Spellbinding Ways to Make Nature More Visible*. Simon & Schuster.

Bellamy Foster J, Clark B, 2020. *The Robbery of Nature: Capitalism and the Ecological Rift*. Monthly Review Press.

Bilgrami A (editor), 2020. *Nature and Value*. Columbia University Press.

Blackmore S, Oldfield S, 2017. *Plant Conservation Science and Practice: The Role of Botanic Gardens*. Cambridge University Press.

Cardinale B, Primack R, Murdoch J, 2019. *Conservation Biology*. Oxford University Press.

Chandler D *et al.* (editors), 2020. *Resilience in the Anthropocene: Governance and Politics at the End of the World*. Routledge.

Clewell AF, Aronson J, 2013. *Ecological Restoration: Principles, Values, and Structure of an Emerging Profession* (2nd edition). Island Press.

Coyle Det al. (editors), 2011. *The Economics of Enough: How to Run the Economy as if the Future Matters*. Princeton University Press.

Dennis R, 2021. *Restoring the Wild: 60 Years of Rewilding Our Skies, Woods and Waterways*. William Collins.

Garner R, 2018. *Environmental Political Thought: Interests, Values and Inclusion*. Red Globe Press.

Gilbert F, Gilbert H, 2019. *Conservation: A people-centred approach*. Oxford University Press.

Gillson L, 2015. *Biodiversity Conservation and Environmental Change: Using Palaeoecology to Manage Dynamic Landscapes in the Anthropocene*. Oxford University Press.

Heywood VH (editor), 1995. *Global Biodiversity Assessment*. Cambridge University Press.

Hill P, 2017. *Environmental Protection: What Everyone Needs to Know*. Oxford University Press.

Holl KD, 2020. *Primer of Ecological Restoration*. Island Press.

IPBES, 2019. *The Global Assessment Report on Biodiversity and Ecosystem Services of the Intergovernmental Science-Policy Platform on Biodiversity and Ecosystem Services*. Brondizio ES, Settele J, Díaz S, Ngo HT (editors). IPBES secretariat. (https://ipbes.net/global-assessment)

Jeffery MI, Firestone J, Bubna-Litic K (editors), 2008. *Biodiversity, Conservation, Law + Livelihoods: Bridging the North–South Divide*. Cambridge University Press.

Jepson P, Blythe C, 2020. *Rewilding: The Radical New Science of Ecological Recovery*. Icon Books.

Kallis G *et al.*, 2020. *The Case for Degrowth*. Polity.

Larkin A, 2013. *Environmental Debt: The Hidden Costs of a Changing Global Economy*. St Martin's Press.

Lockwood M, Worboys G, Kothari A, 2012. *Managing Protected Areas: A Global Guide*. Routledge.

Manning RE *et al.* (editors), 2016. *A Thinking Person's Guide to America's National Parks*. George Braziller Inc.

Martin A, 2017. *Just Conservation: Biodiversity, Wellbeing and Sustainability*. Earthscan Conservation and Development series. Routledge.

McManis CR (editor), 2009. *Biodiversity & the Law: Intellectual Property, Biotechnology & Traditional Knowledge*. Earthscan.

Millennium Ecosystem Assessment, 2005. *Ecosystems and Human Well-being: Biodiversity Synthesis*. World Resources Institute. (https://www.millenniumassessment.org/documents/document.354.aspx.pdf)

Monbiot G, 2016. *How Did We Get Into This Mess? Politics, Equality, Nature*. Verso.

Moore JW (editor), 2016. *Anthropocene or Capitalocene? Nature, History, and the Crisis of Capitalism*. Kairos, PM Press.

Morris J, 2016. *Sustainable Control of Invasive Plants: Adding Value to Biodiversity Economy for Ecosystem Services*. LAP Lampert.

Morrison ML, Mathewson HA (editors), 2015. *Wildlife Habitat Conservation: Concepts, Challenges, and Solutions*. Johns Hopkins University Press.

Primack R, 2014. *Essentials of Conservation Biology* (6th edition). Sinauer.

Root TL, Hall KR, Herzog MP, Howell CA, 2015. *Biodiversity in a Changing Climate: Linking Science and Management in Conservation*. University of California Press.

Surampalli RY *et al.* (editors), 2020. *Sustainability: Fundamentals and Applications*. John Wiley & Sons.

Tisdell CA, 2014. *Human Values and Biodiversity Conservation: The Survival of Wild Species*. Edward Elgar Publishing Ltd.

Vadrot ABM, 2016. *The Politics of Knowledge and Global Biodiversity*. Routledge.

8 No one is too small to make a difference

Bradford L, 2019. *Save the World: There Is No Planet B: Things You Can Do Right Now to Save Our Planet*. Summersdale.

Carson R, 2000. *Silent Spring*. Penguin.

Chillingsworth J, 2019. *Live Green: 52 steps for a more sustainable life*. Quadrille Publishing Ltd.

Olivia M, 2020. *Minimal: How to Simplify Your Life and Live Sustainably*. Ebury Press.

Thunberg G, 2019. *No One Is Too Small to Make a Difference*. Penguin.

Index

References to figures and tables are in *italics*.